TECHNOLOGICAL SYSTEMS AND DEVELOPMENT

Also by Jeffrey James

CONSUMER CHOICE IN THE THIRD WORLD
CONSUMPTION AND DEVELOPMENT
ECONOMICS OF NEW TECHNOLOGY IN DEVELOPING COUNTRIES (*with F. Stewart*)
IMPROVING TRADITIONAL RURAL TECHNOLOGIES
TECHNOLOGY, INSTITUTIONS AND GOVERNMENT POLICIES (*with S. Watanabe*)
THE IMPACT OF WORLD EMPLOYMENT PROGRAM RESEARCH ON TECHNOLOGY (*with F. Stewart, A. Bhalla and others*)
THE STATE, TECHNOLOGY AND INDUSTRIALIZATION IN AFRICA
THE TECHNOLOGICAL BEHAVIOUR OF PUBLIC ENTERPRISES IN DEVELOPING COUNTRIES
THE TRANSITION TO EGALITARIAN DEVELOPMENT (*with K. Griffin*)

Also by Haider A. Khan

DEBT AND DEVELOPMENT IN AFRICA
MACROECONOMIC EFFECTS AND DIFFUSION OF ALTERNATIVE TECHNOLOGIES WITHIN A SOCIAL ACCOUNTING
MATRIX FRAMEWORK: The Case of Indonesia (*with Eric Thorbecke*)
TECHNOLOGY, ENERGY AND DEVELOPMENT: The South Korean Transition
THE POLITICAL ECONOMY OF SANCTIONS AGAINST APARTHEID
TRANSITIONAL ECONOMIES AND REGIONAL ECONOMIC DEVELOPMENT STRATEGIES (*editor with A. Kumssa*)

Technological Systems and Development

Jeffrey James
Professor of Development Economics
Tilburg University
The Netherlands

and

Haider A. Khan
Associate Professor
Graduate School of International Studies
University of Denver
Denver, Colorado
USA

First published in Great Britain 1998 by
MACMILLAN PRESS LTD
Houndmills, Basingstoke, Hampshire RG21 6XS and London
Companies and representatives throughout the world

A catalogue record for this book is available from the British Library.

ISBN 0–333–67023–X

First published in the United States of America 1998 by
ST. MARTIN'S PRESS, INC.,
Scholarly and Reference Division,
175 Fifth Avenue, New York, N.Y. 10010

ISBN 0–312–21074–4

Library of Congress Cataloging-in-Publication Data
James, Jeffrey.
Technological systems and development / Jeffrey James and Haider A. Khan.
p. cm.
Includes bibliographical references and index.
ISBN 0–312–21074–4
1. Technological innovations—Economic aspects—Developing countries. 2. Labor market—Developing countries. 3. Income distribution—Developing countries. I. Khan, Haider. II. Title.
HC59.72.T4J35 1998
338'.064—DC21 97–18335
CIP

This book is printed on paper suitable for recycling and made from fully managed and sustained forest sources.

10 9 8 7 6 5 4 3 2 1
07 06 05 04 03 02 01 00 99 98

Printed and bound in Great Britain by
Antony Rowe Ltd, Chippenham, Wiltshire

Contents

List of Tables

List of Figures

Acknowledgements

Part of the research on which this book is based was undertaken with the assistance of travel grants awarded to Haider Khan by the Dutch National Science Foundation (NWO). Those grants were supplemented by financial support from the Center for Economic Research at Tilburg University. We are grateful to both institutions for giving us such valuable opportunities for collaboration. We are also grateful to Erik Thorbecke for allowing us in Chapters 5 and 6 to draw heavily on the results of his joint research with Haider Khan on Indonesia, which appeared originally in a book entitled *Macroeconomic Effects and Diffusion of Alternative Technologies within a Social Accounting Matrix Framework*, Gower, 1988.

Four of the chapters were originally published in journals. They are reprinted here with the kind permission of the publishers. Chapter 2, written by Jeffrey James, is reprinted from *World Development*, vol. 4, 1976, pp. 143–9, with permission from Elsevier Science Ltd., The Boulevard, Langford Lane, Kidlington 0X5 1GB, UK. Chapters 5 and 6, written jointly by Jeffrey James and Haider Khan, are also reprinted from *World Development*, the former from vol. 21, 1993, pp. 817–27 and the latter from vol. 25, 1997, pp. 153–65, with permission from Elsevier Science Ltd., The Boulevard, Langford Lane, Kidlington OX5 1GB, UK. Chapter 3, written by Jeffrey James, is reprinted from the *Journal of Development Economics*, vol. 7, 1980, pp. 175–89 with kind permission from Elsevier-NL. Sara Burgerhartstraat 25, 1055 KV Amsterdam, The Netherlands.

The authors and publishers also wish to thank the following who have kindly given permission for the use of copyright material: Oxford University Press for the table on p.147 of H. Chenery et al., *Redistribution with Growth*, 1976, by permission of the publisher; the International Labour Office for the table on p. 40 of *Appropriate Products, Employment and Technology. Case-Studies on Consumer Choice and Basic Needs in Developing Countries*, edited by W. van Ginneken and C. Baron, 1984, copyright © International Labour Organization, 1984. Published on behalf of the ILO by Macmillan Press Ltd. London; Dr H. Sandee for the table on p. 62 of his PhD thesis *Innovation Adoption in Rural Industry. Technological Change in Roof Tile Clusters in Central Java, Indonesia*, Free University, Amsterdam, 1995; Elsevier Science for the figure on p. 218 of A. Amsden, 'The Division of Labour is Limited by the Type of Market: the Case of the Taiwanese Machine Tool Industry', *World Development*, vol. 5(3), 1977, pp. 217–33, with permission from Elsevier Science Ltd., The Boulevard, Langford Lane, Kidlington, Oxford OX5 1GB, UK; Macmillan Press Ltd for tables on pages 2 and 270 of F. Stewart, *Technology and Underdevelopment*, 1977; Ashgate Publishing Limited for tables on pages 45, 46, 68 and 69 of H. Khan and E. Thorbecke, *Macroeconomic Effects and Diffusion of Alternative Technologies within a Social Accounting Matrix Framework*, 1988.

We are grateful, finally, to Corina Maas for her prompt and accurate typing assistance, and to Anne Rafique for expert and thorough editorial assistance.

1 Technological Systems in Developing Countries

Drawing heavily on ideas presented originally by Stewart[1] this book will explore the notion of a technological system and show how this concept can be applied empirically to a range of key policy issues in development economics. We will argue that although technological systems are rooted in very different historical periods, they can nevertheless still be identified as distinctive entities in the context of today's developing countries. As such our argument has two main components, one being historical in character and the other having to do with the contemporary relationships between different technological systems in developing countries. Before presenting our ideas on these two issues, however, we need to investigate the concept of technology itself.

THE CONCEPT OF TECHNOLOGY

Following Stewart, technology can be said to comprise a series of techniques, each of which, in turn, consists of a set of characteristics.[2]

> In discussing 'characteristics' of technology, one should include all the relevant features which determine its resource use, productivity, and impact on production and consumption patterns. These features include the nature and design of the product, the scale and organisational system for which the technology is designed, its resource use, including capital and labour intensity, materials and fuel use, skill requirements, and the infrastructural and complementary inputs it requires.[3]

Technology can thus be depicted in matrix form, with each column describing the characteristics associated with each technique, as shown in Table 1.1.

Our concern in this book is with the relationships between these characteristics of the different techniques that together make up the technology in existence in the developing countries. What we want to show are patterns in the combinations of characteristics within and

Table 1.1. The concept of technology

Techniques / *Characteristics*	1	....	n
Organization of production			
Product type			
Product nature			
Scale of production			
Material inputs			
Labour input:			
skilled			
unskilled			
Capital inputs			
Location of production			
Managerial inputs			
Infrastructural requirements			

Source: Stewart (1977).

between techniques that form distinct technology systems. Traditional economics, we should emphasize, provides very little help with this endeavour in that it tends to focus largely on just one technical characteristic, namely, factor intensity.[4] This tendency is readily apparent from the standard diagram in the choice of technology literature, which portrays the issue purely in terms of a range of capital–labour ratios whose costs vary according to the relative prices of the two inputs. The same tendency is also apparent, however, in the way that the direction of technical change is conceived in traditional theory. The theory of induced technological innovation, for example, 'essentially postulates that changes in relative factor scarcities are reflected in changes in relative factor prices which in turn guide technological progress towards saving on the factors that become relatively more expensive'.[5]

A more useful point of departure for our purposes historically is the Marx-Schumpeter (and later, as we shall see, Keynes) approach, partly because it encompasses more of the characteristics that are associated with different techniques and partly because it helps to explain the historical basis of the relationships between those characteristics.

THE HISTORICAL GENESIS OF TECHNOLOGICAL SYSTEMS

The first of these features of Marx's approach rests on two concepts that jointly define his view of historical materialism, namely, the forces and the relations of production. Embracing as it does, labour, capital and raw materials,[6] the former concept corresponds broadly to the input characteristics listed previously in Table 1.1. The second concept, on the other hand, refers more to the organizational characteristics mentioned in that table. For what Marx seemed to mean by 'relations of production' was 'EITHER relations of ownership by persons of productive forces or persons OR relations presupposing such relations of ownership'.[7] Thus the producer may or may not own his labour-power and he may or may not own his means of production. The case where both factors are owned by the producer, for example, corresponds to a situation where independent producers own their means of production which are operated without the assistance of wage labour. Under capitalism, on the other hand, the producer owns his means of production which is operated on the basis of wage labour.

Modes of production

In Marxism, the relationships between the forces and relation of production are analysed within distinct modes of production.

> The mode of production is an internally balanced whole in which production relations, and especially the base on which they rest–the ownership of the means of production–are adjusted to the requirements of a given state of the development of social productive forces. We know that production relations cannot be arbitrary for given productive forces.[8]

An obvious but nonetheless important example of the way in which the relations of production conform to the forces, is provided in Chapter 6 below. The case in point there is a plant such as an iron or steel works, which, because it employs highly specialized equipment and produces on a large scale, cannot be established on the basis of simple commodity production, with large numbers of independent producers. Such technology needs to be established instead on the basis of capitalist or socialist relations of production.[9] Nor, to take another example, 'was the development of large-scale industry based on modern technology possible under feudalism since such development is dependent on the emancipation of the peasants from serfdom and the emergence of a class of free hired workers who can be employed in large factories'.[10] Still another example of the same point can be made in relation to slavery which, 'could not be the general condition of producers in a society of computer technology, if only because the degree of culture needed in labourers who can work that technology would lead them to revolt, successfully, against slave status'.[11]

Schumpeter is perhaps the greatest economist since Marx to have attempted an analysis of technological change and innovation. Starting with his *Theory of Economic Development*, Schumpeter gradually made innovation the central factor in economic development. Recently Futia has presented a formal theory of Schumpeterian innovation. In traditional technological systems innovation is at best haphazard and without any serious economic consequences. The capitalist system of technology, however, proceeds through 'creative destruction'.

As Goodwin has pointed out, it is actually possible to identify a Marx-Keynes-Schumpeter (MKS) approach to growth and capital accumulation under capitalism.[12] According to Goodwin, who laments his lack of success in making Schumpeter appreciate Keynes or learn the linear differential system of cycles, there is a connection between the long-run view of Schumpeter and the short-run view of Keynes. Recent work by Geroski and others has connected the demand-induced effects via business cycles on innovation.

One class of models we present and apply in the latter parts of the book can be seen as Keynesian models rooted in a generalized Marx-Schumpeter conception of technology systems. Thus the social accounting matrix based fixed price models can be seen as directly inspired by the MKS approach in Goodwin's terminology. This is also consistent with, and in fact complementary to, a number of neo-Schumpeterian approaches that have sprouted recently.[13]

According to our approach, therefore, capitalism tends to be associated with a rather different set of technical characteristics than pre-capitalist modes of production, in terms of inputs, organization and scale of production. On the product side, too, as a number of the following chapters point out, there is a fundamental distinction between pre-capitalist and capitalist modes of production. For, whereas in the former 'the greater part of the products are produced for the immediate needs of the community not as commodities',[14] the latter turns 'the commodities formerly produced as immediate use-values into exchange values'.[15]

Since contemporary developing countries are characterized by the coexistence of modes of production which originate in entirely different historical periods, the distinctions drawn in the previous paragraph (and summarized in Table 1.2), form the cornerstone of our analysis of

Table 1.2 A schematic comparison of capitalist versus non-capitalist modes of production

Mode of production / *Nature of production*	*Pre-capitalist*	*Capitalist*
Relations of production	Non-wage labour (e.g. independent producers, feudalism)	Wage-labour, means of production owned by producer
Forces of production	Relatively low level of productive power	Relatively high level of productive power
Scale of production	Economies of scale unimportant	Economies of scale often important
Purpose of production	Use value (for satisfaction of the immediate needs of the community)	Exchange value

technological systems. Our main concern in the chapters below, however, is not so much with capitalism in general as it is with the particular type of capitalism that has emerged in today's developed countries. For it is the technological system belonging to this form of late capitalism that is imported by and dominates much of the industrial sector in contemporary developing countries.

The type of society that is our primary concern is thus one where a modern technological system imported from the developed countries, coexists with an indigenous technological system whose characteristics are essentially pre-capitalist in origin. This type of society, we should note, exhibits much of the economic and social dualism described in Boeke's influential early work on the subject.[16] For what interested him was not simply a society where two or more distinct social systems coexist. Rather, he was concerned with the type of dualism where 'one of the two prevailing social systems, as a matter of fact always the most advanced, will have been *imported from abroad* and have gained its existence in the new environment without being able to oust or to assimilate the divergent social system that has grown up there... Without doubt the most frequent form of social dualism is to be found where an imported western capitalism has penetrated into a precapitalistic agrarian community'.[17] Social dualism is thus defined as 'the clashing of an imported social system with an indigenous social system of another style'.[18]

As Boeke did, we shall have to pay careful attention to the interrelationships between these two components of the dualistic society. If one is adequately to describe the nature of those relationships, however, it is essential to understand – at least in broad terms – the historical evolution of the technological system under capitalism in the developed countries and it is accordingly to this issue that we turn next.

The evolution of technological characteristics in the developed countries

The theory of induced innovation mentioned earlier can account for part of the change in the characteristics of technologies in the developed countries over time. In particular, the increase in capital relative to labour and the resulting rise in the relative price of the latter, has resulted in a general bias towards labour-saving innovations (as reflected in a steady rise in the capital to labour ratio over time in the developed countries). What this theory is not capable of explaining, however, are the changes in some of the other characteristics of techniques mentioned in connection with Table 1.1. Products, for example, have changed dramatically during this century, as Stewart in particular has emphasized. As she sees it,

> Technical change in products in developed countries occurs in line with the rising incomes (and is partly responsible for those rising incomes), so that the balance of characteristics offered by new products corresponds to the changing demands of consumers as their incomes rise. The rising incomes have the effect of shifting demand towards different products with more sophisticated, labour-saving, higher-quality, etc. characteristics: to summarise this complex of changes we may say that the characteristics of the new products have more high-income characteristics.[19]

Stewart also shows how, via a complex series of mutually interactive relationships between income and other variables, a number of the other major characteristics of technology have changed in systematic fashion. In general, one can say – with a good deal of empirical backing – that there has been an increase in the scale of production, an increase in the ratio of skilled to unskilled workers and in the degree of sophistication of managerial methods, and a rise in the productivity of labour.[20]

The dynamic process she describes turns heavily on the notion of technology as a *package* 'in which each technique has to fit into the whole.... the efficiency or otherwise of a particular technique depends not only on its own performance, and that of its immediate substitutes, but also on the surrounding technology'.[21] New techniques, accordingly, are continuously being developed against the background of successive packages: that is, particular sets of inputs, infrastructure, legal and administrative systems, labour skills and so on.

What developing countries tend to import from the industrialized world, therefore, is a technology *package* rather than simply a single technique, since each such technique 'imposes requirements for linked techniques'.[22] And this in turn means that developing countries often have to import those linked techniques as well, if the original technique is to be viably operated. Thus it is that one can speak of the importation of an entire modern technological system from developed to developing countries.

THE COEXISTENCE OF TECHNOLOGICAL SYSTEMS

It is one thing to argue, as we have, that the characteristics of techniques differ systematically according to modes of production and the historical evolution of capitalism in the developed countries. It is quite another thing, however, to assert that these historically based patterns will retain their characteristics in the contemporary developing country context. For when the various modes of production (and their related technological systems) coexist in this context, the possibility arises that they will interact in ways that blur or entirely reverse the historically derived patterns.

What is most relevant for our purposes is the possibility that the modern, imported technological system will provide inputs and final goods to the traditional, pre-capitalist technological system or that it will purchase inputs and final goods from that system. In fact, as Table 1.3 shows, examples of each such possibility can readily be found in the Third World.

The question at issue, however, is whether these are merely isolated examples or part of a pervasive phenomenon in the Third World.[23] As we see it, there are two main reasons for thinking that the former description is much closer to the truth than the latter. One of them concerns the first row of Table 1.3 (that is, the likelihood that the modern technology system utilizes the products of the traditional system) and the other concerns the second row of that table (that is, the likelihood that the modern technology system will supply goods to the traditional system).

Table 1.3 Interrelationships between coexisting technological systems in developing countries

	Inputs	*Outputs*
Goods supplied by traditional to modern technology system	Subcontracting labour-intensive components of textile production to self-employed artisans	Luxury products consumed by high-income, urban households
Goods supplied by modern to traditional technology system	Use of low-cost plastic materials by labour-intensive artisanal footwear producers	Consumption by low income rural households of modern products, such as infant formula, soft drinks, etc.

The reason for supposing that the modern technological system will only be reliant to an incidental degree on inputs produced by traditional technologies is mainly a technical one. It has to do with the increasing extent to which modern technologies impose physical requirements on the rest of the production system. In particular, to quote Stewart again,

> It is on the whole a characteristic of more recently developed techniques that the room for variation in associated techniques has declined. This has occurred partly as a result of the more sophisticated scientific techniques involved, so exact temperature control, etc., is required, and also with mechanisation and then automation. *Mechanical methods impose more rigorous requirements than hand methods because hand methods may correct for variations in previous processes. ... Automated techniques leave even less room for human correction of previous variability – and hence impose strict requirements of standardisation and detailed specification on previous processes. The effect is that a given automated technique is designed to be operated using the output of certain specified techniques for previous and subsequent stages and variation may not be possible. Thus the links between different parts of the system have got stronger.*[24]

Thus it is that the products of traditional technologies tend to be excluded by the demands of the modern technological system. Exceptions to this general pattern do occur, however, perhaps most typically where parts of the production process have remained resistant to automation. This is the case, for example, with certain phases (e.g. assembly) in the clothing industry, which explains the phenomenon noted in Table 1.3, of subcontracting to traditional artisans by large-scale modern producers. (In the international context this same phenomenon manifests itself as subcontracting by multinationals of labour-intensive operations to low-cost locations in the Third World.)[25]

Looked at from the opposite point of view – of the modern technological system as supplier rather than purchaser of inputs and final goods – the previous quotation suggests that there might be more scope for exchange, since, as noted there, traditional methods of production are more flexible in their input demands. Those methods might therefore be expected to use a more varied range of inputs, produced by techniques of different vintages and capital intensities. Against this, however, has to be set the fact that the products of the modern technological system are designed for high rather than low incomes and as such are unlikely to be extensively demanded by the traditional technology system.

To an important degree, this offsetting influence operates through multinational corporations, since it is these institutions that dominate the modern technological system of many developing countries.[26] There are of course cases where these firms specifically adapt their products to the needs of low-income consumers (as, for example, with certain types of specially fortified foods).[27] Measured against the total activity of foreign-owned firms in developing countries, however, such cases are few, mainly because these firms (as is often observed) tend to emphasize product differentiation rather than price competition and this effectively limits the degree to which product adaptations are accessible to the poor.[28] The problem can be illustrated with the help of Figure 1.1, which shows the costs of producing, and the demand for, the type of product that is typically sold by the multinational firm in a developing country.

Given its 'high-income' features, including the emphasis on brand differentiation (which usually involves heavy outlays on packaging, advertising and marketing), the costs of production, we assume, are represented by LRAC1. At these costs and prices, 'high-income' demand, represented by D1 can be met but the demand curve representing the poor majority (D2) lies entirely below the cost of production. The 'low-income' demand in this case can be met only by a lower-cost product that embodies a higher proportion of essential characteristics, as represented for example by LRAC2. Such a product, we have suggested, will tend to emanate from the traditional rather than the modern technology system.[29]

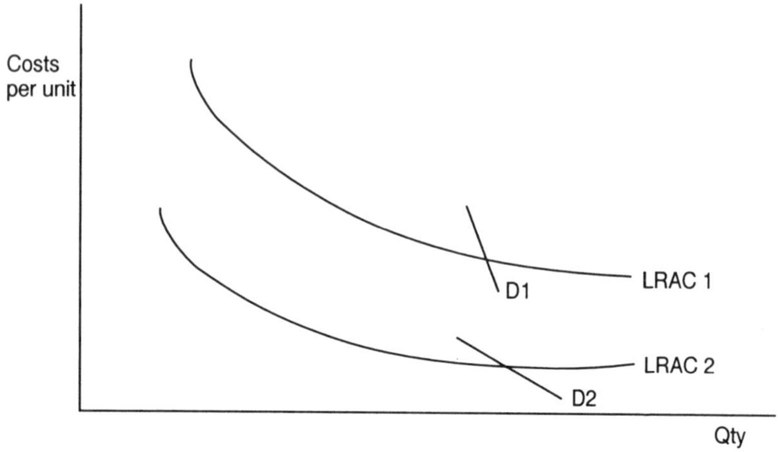

Figure 1.1 Product characteristics and the type of market
Source: Based on Amsden (1977).

CONSOLIDATION OF THE ANALYSIS AND AN EXAMPLE

In developing the notion of a technological system, we have relied first on the historical evolution of modes of production and second on the manner in which these various modes coexist in the context of today's developing countries. By combining these two aspects of our analysis we can now define more fully the two polar types of technological systems with which we shall mainly be concerned, namely, the traditional and the modern. The distinguishing features of these two types of systems are juxtaposed in Table 1.4.

Most of these distinctions can usefully be illustrated by a case study of building materials in Kenya, which was undertaken by Frances Stewart.[30] As shown in Table 1.5, the use of these materials in that country varies from mud and wattle at the one extreme, to cement blocks made with a chemical additive at the other. It can also be seen from Table 1.5 that there is a fairly clear association between the characteristics of these various products and the factor intensity of the technology that is used to produce them. In particular, the 'higher-income' products tend to be associated with the more capital-intensive techniques, although stone production, a high-income product, is manufactured in a labour-intensive manner. Stewart's interpretation of this association is very much in line with our earlier analysis of how traditional and modern technological systems have evolved historically. Thus,

Traditional methods (mud and wattle, and sun-dried blocks) were developed in societies where labour and materials were the only resources available so naturally they used only these resources; and they were developed for low-income consumers. In contrast, such techniques as pre-cast units with foaming additives, recently developed in advanced countries for consumption there, tend to be relatively investment-intensive and with high-quality characteristics for high-income consumers.[31]

Table 1.4 Traditional versus modern technology systems: a stylized comparison

Traditional technology system	*Modern technology system*
Labour-intensive methods of production	Capital-intensive methods of production
Production for 'use-value' involving 'low-income' products	Production for 'exchange value' involving 'high-income' products
Non-capitalist mode of production (especially independent producers, family labour)	Capitalist mode of production, based heavily on multinational corporations
Dispersed small-scale (often rural) production units	Production concentrated in small number of large-scale units (in urban areas)
Relatively high labour–output ratio throughout production chain	Relatively low labour–output ratio throughout production chain
Relatively high non-labour (input) to output ratio throughout production chain	Relatively low non-labour (input) to output ratio throughout production chain
Uses mainly inputs from own system	Uses mainly inputs from own system (often in the form of inputs from developed countries)
Consumes a relatively high proportion of its own products	Consumes a relatively high proportion of its own products

In two other respects as well, the relationships depicted in Table 1.5 conform closely to the stylized patterns described in Table 1.4. The first is that – as hypothesized in that latter table – the traditional and modern technological systems tend to rely on their own, rather than each other's inputs of basic materials (see, for example, the distinction between the materials used for mud and wattle constructions and those

Table 1.5 Alternative technological systems in the Kenyan cement block industry

Material	*Quality of building*	*Income class of consumer*[1]	*Source of basic material*[2]		*Labour intensity*[3]		
			Place	*Economic sector*	*Material production*	*Processing*	*Building*
Mud and wattle	low	subsistence	local	informal	high	high	high
Sun-dried clay blocks	low to medium	slightly above subsistence (rural) low (urban)	local	informal	high	high	high
Murram-enforced blocks	low to medium	slightly above subsistence (rural) low (urban)	local	informal formal[4]	high	high-medium	high
Black cotton bricks	low to medium	slightly above subsistence (rural) low (urban)	local	informal	high	high	high
Stones	high	mainly high	local	informal formal	high	high	high
Timber	medium	medium and high	local	informal formal	[5]	[5]	[6]
Pre-cast concrete panels	medium	medium and high	local	formal	medium[7]	medium[7]	medium
Pre-cast concrete panels made with a foaming agent	medium	medium and high	local and imported	formal	medium[7]	medium[7]	low-medium
Cement blocks made with a chemical additive	high	medium and high	local and imported	formal	low	low	low-medium
Cement blocks	medium	medium and high	local	formal	medium[7]	medium[7]	medium

1 Low: rural subsistence and most small-scale market production, urban unemployed, under-employed – including most informal-sector activities. Medium: unskilled and semi-skilled formal-sector (mostly urban) workers. High: skilled and professional. The use of this classification means that the majority of the population should be classified as low-income.

2 This refers both to geographical location (local or imported) and to sector of the local economy, which is regarded as being divided into an informal and a formal sector. Broadly, the informal sector is more labour-using, while formal-sector production involves more equipment, often imported.

3 To classify production methods according to labour intensity implies that each material may be identified with a single technique; as the survey of cement block making showed, that is not the case, and considerable variation may be possible.

4 Cement.

5 Intensive use of skilled labour and natural resources.

6 Intensive use of skilled labour and some use of machinery.

7 Varies (see footnote 3).

Source: Stewart (1977).

used in the making of cement blocks with a chemical additive). The second point that needs to be emphasized is that the relationship between factor intensity, product characteristics and incomes applies not just to one phase in the production process, but rather to all such phases, again illustrating the systems perspective adopted in this chapter and summarized in Table 1.4. Thus, just as the traditional techniques use labour-intensive methods in material production, processing and building, so too do the modern techniques tend to use the same (more capital-intensive methods) at each of those stages in the production process. As Stewart puts it, 'Later techniques from advanced countries are designed to be operated with other techniques and products of the same vintage,'[32] and these linkages between different techniques put the traditional sector at a disadvantage in supplying products and techniques to the modern sector.

Methodology

In the empirical chapters below we examine various different facets of technological systems and we also use these analyses to study a range of problems in development economics, such as the choice of technique, the impact of alternative technologies on the distribution of income and the employment effects of an income redistribution. We cannot therefore properly speak of any single methodology that is uniformly applied throughout the book as a whole. There is, however, one methodological issue that is common to all the empirical applications of the notion of a technological system and for this reason it warrants our attention in this introductory chapter. This common factor has to do with the requirements for the identification of technological systems.

Aggregation and the Invisibility of Technological Systems Because of the way in which data on consumption and production are normally collected in developing countries, the divergent technological systems that were described above are frequently difficult to identify in practice (hence the reference to 'invisibility' in the title of this section). On the consumption side, the problem originates to a large degree in the way that commodities are conceptualized in traditional neoclassical theory. In particular, goods and services tend to be viewed according to this theory as being of a standard or homogeneous quality.[33] This (plainly invalid) assumption has created a major problem in identifying technological systems because it has generally been carried over into the practice of data collection, which, by aggregating products into single homogeneous categories effectively ignores the differences in characteristics between them. (Such differences, as we have seen, are crucial distinctive features of alternative technology systems.)

Thus in many consumer expenditure surveys in developing countries, for example, one finds categories such as 'transport' and 'housing', which are far too broad to capture the multifarious ways in which these needs can be and in fact are met by different categories of consumers. The need for transport can be met by bicycles or automobiles, while the 'need for accommodation may be fulfilled more or less well, by the sidewalks of Calcutta, caves, mud-huts, multi-storey apartments, single family houses, or a palace'.[34] Even a much more narrowly defined product such as rice often embodies very large differences in characteristics such as whiteness, uniformity of grains, cleanliness, consistency and nutrition. Yet, the vast majority of studies on demand tend to treat rice as a uniform, homogeneous product, as Tabor has indicated for the case of Indonesia.[35]

On the production side too, much of the available data is of too aggregated a form to be helpful in the identification of distinct technological systems. Many important data sources for example (such as censuses), tend to exclude firms below a certain minimum size[36] and, partly as a result, it is still the case that although small-scale industry employs a large proportion of the industrial labour-force in many developing countries, 'remarkably little is known about its composition and characteristics'.[37] This, in turn, makes it extremely difficult to associate the production side of traditional technological systems with the corresponding consumption dimensions of those systems, even when the latter have been identified.

Disaggregation On both the consumption and production sides it is evident that the study of technological systems requires a far greater degree of disaggregation than is typical of most available data in developing countries. The exact extent of disaggregation that is required for this purpose, however, depends on which particular aspects of those systems one wishes to investigate. It is useful in this regard to distinguish between two types of disaggregation: disaggregation at the micro level and disaggregation at the macro level. (These levels correspond in fact to the manner in which the following five chapters are divided).

Micro-level Disaggregation In this case, one is concerned with only a subset of the economic relationships that comprise a technological system. What is perhaps most typical of this form of disaggregation is the 'choice of technique study', which begins by defining different products (within a related category) according to the characteristics they embody and then investigates the attributes of the consumers and producers of those products.

Consider, for example, the case of rice that was referred to earlier as a product whose characteristics vary from one brand to another. Each brand will differ, that is to say, according to its whiteness, uniformity of grains, consistency, cleanliness and nutritional attributes. These differences can be objectively measured and the brands positioned on this basis in characteristics space, such as the two dimensions represented in Figure 1.2 (Brand X, for example, is shown as a 'high-income' product because it may have undergone the more intensive processing required for a high degree of whiteness, consistency, cleanliness, etc. Brand Y, on the other hand, scores less well on these characteristics but is produced and sold at a lower cost). One would then need to assess the attributes of the consumers (e.g. income, location) and producers (e.g. method and scale of production, location) of the

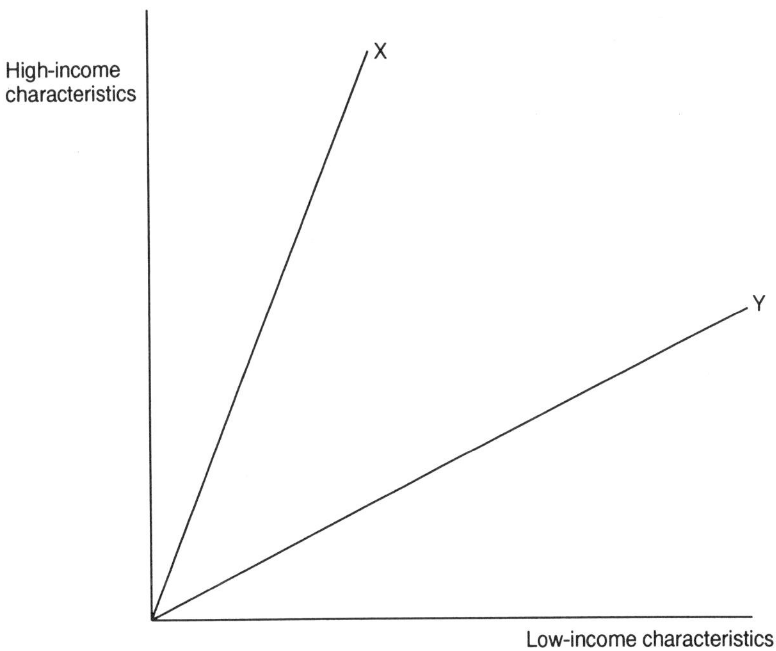

Figure 1.2 High- versus low-income products

different brands thus classified. According to the view of technological systems that was espoused earlier in this chapter, these combined disaggregated data would reveal certain distinct patterns of association. (See Chapters 2 and 3 below).

Macro-level Disaggregation In contrast to the limited extent of disaggregation that is required by the previous example, disaggregation at the macro level is needed to examine the technological system in its entirety. Whereas, for example, the micro choice of technique studies do not concern themselves with the various types of linkage in the economy, the nature and extent of these are central to the economy-wide analysis of technological systems. An intuitive grasp of the type of disaggregation that is needed at this level of analysis can be gained from the simple 'transactions tableau' shown in Table 1.6.

As in the micro-level analysis, there is disaggregation in this table by products, households and firm size. But the difference is that these disaggregated categories are here being used to assess the extent to which producing units in the different sectors are linked to one another and to the various types of consumers in the economy. In several of the chapters in the second part of the book this type of analysis is in fact conducted within the framework of a highly disaggregated social accounting matrix (SAM), that is capable of capturing a large degree of the complexity that underlies the relationships between different technological systems.

Table 1.6 A schematic transactions matrix

Purchasing sector	Goods (as embodied characteristics)			Services (as embodied characteristics)			Households Government		
Supplying sector	Large scale	Small scale	Informal	Large scale	Small scale	Informal	Rich	Poor	
Goods[1]:									
Large scale	X_{11}	–	–	–	–	X_{16}	H_1^1	H_1^2	G_1
Small scale	X_{21}					:	H_2^1	H_2^2	G_2
Informal	X_{31}					:	H_3^1	H_3^2	G_3
Services[1]:									
Large scale	X_{41}	–	–	–	–	:	H_4^1	H_4^2	G_4
Small scale	X_{51}					:	H_5^1	H_5^2	G_5
Informal scale	X_{61}					X_{66}	H_6^1	H_6^2	G_6

[1] Defined as in columns
Source: Rao (1976).

Notes

1. Stewart (1977).
2. Ibid.
3. Stewart (1981) p. 92.
4. This point is central to Stewart's (1977) critique of the neoclassical approach to technology choice.
5. Ahmed and Ruttan (1988) p. 2.
6. Cohen (1978).
7. Cohen (1978) p. 34–5.
8. Lange (1973) p. 21. Opinions differ, however, on the nature of the causality between the forces and relations of production. According to Cohen's well-known interpretation, for example, the latter are explained by the level of development of the former (Cohen, 1978). On the other hand, it is also likely that the relations of production exert some influence over the generation of technology (Richards, 1986), so that in reality causality runs both ways.
9. Lange (1963).
10. Lange (1963) pp. 21–2.
11. Cohen (1978) p. 158, who also cites a number of other reasons 'for thinking that slavery and sophisticated, improving productive forces exclude one another' (p. 10).
12. Goodwin (1991).
13. See for example, Richard Day and Gunner Eliasson (1986) and Frederic Scherer and Mark Perlman (1992); among others for recent neo-Schumpeterian approaches. However, many of these studies ignore demand side as well as distributional issues.
14. Marx (1946) p. 337.
15. Marx (1964) p. 116.
16. Boeke (1953).
17. Ibid., p. 4, emphasis in original.
18. Ibid.
19. Stewart (1977) pp. 17–18.
20. For a detailed discussion see Stewart (1977).
21. Ibid, pp. 8 and 21.
22. Ibid, p. 82.
23. For some authors, examples such as these cast doubt on the existence of a general relationship between product characteristics, incomes and factor intensity. See for example Cooper (1990). As will become evident from our subsequent discussion, we believe instead that these are simply exceptions 'that prove the rule'.
24. Stewart (1977) p. 82, emphasis added.
25. Sharpston (1975).
26. Particularly in Latin America and parts of Asia.
27. For a general discussion of product adaptations by multinationals in developing countries, see James (1993).
28. Ibid.
29. Amsden (1977) uses this diagram to illustrate how Taiwanese machine tool producers were able, by means of lower quality, less

costly products, to penetrate markets that foreign firms found 'impenetrable'.
30. Stewart (1977).
31. Ibid, pp. 269 and 272.
32. Ibid, p. 88.
33. Tabor (1989).
34. Morawetz (1974) p. 513.
35. Tabor (1989).
36. Bell and Duloy (1976).
37. Morawetz (1974) p. 525.

References

I. Ahmed and V. Ruttan (1988). 'Introduction', in I. Ahmed and V. Ruttan (eds),*Generation and Diffusion of Agricultural Innovations: The Role of Institutional Factors*, Aldershot: Gower.

A. Amsden (1977). 'The Division of Labour is Limited by the Type of Market: the Case of the Taiwanese Machine Tool Industry', *World Development*, vol. 5(3).

C. Bell and J. Duloy (1976). 'Statistical Priorities', in H. Chenery et al., *Redistribution with Growth*, Oxford University Press.

J.H. Boeke (1953). *Economics and Economic Policy of Dual Societies*, Haarlem: Tjeenk Willink and Zoon.

G.A. Cohen (1978). *Karl Marx's Theory of History: A Defence*, Oxford: Clarendon Press.

C. Cooper (1990). 'Notes on Income Distribution, Growth and Demand', in F. Stewart, H. Thomas and T. de Wilde (eds), *The Other Policy*, London: IT Publications.

R. Day and G. Eliasson, eds (1986). *The Dynamics of Market Economics*. New York: North-Holland.

R. Goodwin (1991). *Economic Evolution, Chaotic Dynamics and the Marx-Keynes-Schumpeter System*, in Geoffrey M. Hodgson and Ernesto Serpenti, eds, *Rethinking Economics: Markets, Technology and Economic Evolution*. Aldershot: Edward Elgar.

J. James (1993). *Consumption and Development*, London: Macmillan.

O. Lange (1963). *Political Economy*, vol. 1, Oxford: Pergamon.

K. Marx (1946). *Capital*, London: Allen & Unwin.

K. Marx (1964). 'Pre-Capitalist Economic Formations', in E. Hobsbawm (ed.), *Pre-Capitalist Economic Formations*, London: Lawrence & Wishart.

D. Morawetz (1974). 'Employment Implications of Industrialisation in Developing Countries: A Survey', *Economic Journal*, vol. 84.

D.C. Rao (1976). 'Urban Target Groups', in H. Chenery et al., *Redistribution with Growth*, Oxford University Press.

F. Scherer and M. Perlman (1992). *Entrepreneurship, Technological Innovation, and Economic Growth: Studies in the Schumpeterian Tradition*. Ann Arbor: University of Michigan Press.

M. Sharpston (1975). 'International Subcontracting', *Oxford Economic Papers*, vol. 27.

F. Stewart (1977). *Technology and Underdevelopment*, London: Macmillan.

F. Stewart (1981). 'International Technology Transfer: Issues and Policy Options' in P. Streeten and R. Jolly (eds) *Recent Issues in World Development*, Oxford: Pergamon.

S. Tabor (1989). *Price and Quality of Rice in Java: An Investigation into the Demand for Closely Related Goods*. Unpublished PhD dissertation, Free University, Amsterdam.

Part One
Micro Aspects and Applications

2 Products, Processes and Incomes: Cotton Clothing in India

Developing countries are characterized by the coexistence of forms of production which originate in entirely different historical periods. A significant proportion of the production structures of these economies comprises production modes which belong to the period of pre-capitalist economic formations. In this period, as noted in the previous chapter, 'the greater part of the products are produced for the satisfaction of the immediate needs of the community not as commodities',[1] in contrast to the later capitalist mode of production which turns 'the commodities formerly produced as immediate use-values into exchange values'.[2] The genesis of the relationship between products, processes and incomes thus lies in communities where the production mode exists to provide essential commodities for the subsistence requirements of the community. The survival of these earlier modes in developing countries reflects the fact that vast numbers still live at the subsistence level of income and thus require the type of commodity appropriate to that of the earlier period. The organized sector of these economies by contrast operates on the basis of imported processes of production developed for the incomes and product requirements of advanced economies.[3] This chapter attempts a cross-sectional analysis of the relationship between the processes of different modes of production, the nature of commodities produced and the income levels of consumers in the Indian cotton clothing industry. The cross-section results are interpreted in the light of the historical dynamics of technology in the industry.

THE INDIAN COTTON TEXTILE INDUSTRY

The cotton industry in India was chosen to investigate the relationship for a number of reasons:

(a) India has a large proportion (44.5 per cent) of the population below the absolute poverty level of 50 dollars per head.[4] There is consequently a low proportion of wage employment in the economy, and this proportion is lowest in the clothing and footwear industries.[5]

(b) The processes of production adopted by the non-wage sector are significantly different from those of the organized sector.

(c) The quality of cloth produced by the different sectors provides a useful way of defining the essential intensiveness of the respective commodities.

(d) Data are available on both the production and consumption aspects of the industry.

Development of the Production Structure

The spread of commerce 'gave weaving a quantitative and qualitative stimulus which wrenched it out of the form of production hitherto existing'[6] in terms of which production by the peasants for their own use was joined by production for the market. The introduction of machinery gave birth to the mill industry which concentrated in areas near the cotton-growing tracts, near transport, land and power facilities. Some idea of the extent to which the mill industry has developed at the expense of the hand-loom industry can be obtained by considering that the former's share in total output has increased from 43 per cent to 62 per cent between 1909 and the present.

Currently, the industry is divided into three producing groups, namely hand-loom, power-loom and mill producers. The former are invariably self-employed and the latter hire workers on a wage basis. Power-loom producers may operate on a self-employed basis or by employing a fair number of workers under the same shed. The decentralized sector, as it is called (comprising hand-looms and power-looms), produces some 47 per cent of the total output of the industry.[7] Of this, the share of hand-looms is 62.5 per cent. The power-loom sector is in a sense intermediate between hand-looms and mills, comprising as it does mainly small units of four looms and less, with only 5 per cent of the units defined as large in terms of the Industries Act.[8]

Process of Production between Sectors

The processes of production between the sectors are set out in Table 2.1. It is important for what follows to establish the extent of auto-

Table 2.1 Processes of production in the hand-loom and mill sectors

Process	*Form of employment*	*Fixed capital per unit of production, rupees*	*Working capital per unit of production, rupees*	*Total productive capital per unit of production, rupees*	*Total productive capital per employed person*
Hand-loom sector					
1. Throwshuttle hand-loom	self-employment	5	300	305	243.3
2. Flyshuttle hand-loom	self-employment	40	450	490	392
3. Banaras semi-automatic hand-loom	self-employment	200	1,500	1,700	1,133.3
4. Madanpura semi-automatic hand-loom	self-employment	250	2,500	2,500	1,666.6
Mill sector					
1. Mill non-automatic powerloom	wage	4,000	6,225	10,225	20,450
2. Mill automatic powerloom	wage	10,000	7,500	17,500	280,000

Source: Prasad (1963), table 80.

matic loomage, the import content of the production processes and their age structure.

As a percentage of the total of looms installed in the mills only 16.3 per cent are automatic.[9] This figure may be compared with the comparable figures for Germany, Japan and the United States of 69, 81 and 100 per cent respectively.[10]

While hand-looms are always produced domestically,[11] of the total looms in mills 76 per cent are imported.[12] But the age structure of the looms implies that the machinery currently in use is that which was designed for much earlier periods in the exporting countries.[13] The average age-composition of the looms is shown in Table 2.2.

Table 2.2 Age-composition of looms

Pre-1910	*1910–25*	*Post-1925*
57.3	20.7	28.0

Sources: Kumar, Nag and Venkataraman (1965), p. 93.

The differential productivities and labour requirements per loom are shown in Table 2.3 for four of the five processes shown in Table 2.1.

Having established the extent and nature of the respective production processes we turn now to an examination of the nature of the products produced and thereafter to the income classes of consumers of such products.

Table 2.3 Output per loom and labour (shift) per loom

	Output per loom per day	*Labour per loom per shift*	*Number of shifts per day*	*Labour shift loom per day*
Flyshuttle hand-loom	6	1	1	1
Banaras semi-automatic hand-loom	20	1	1	1
Mill non-automatic power-loom	80	1/2	2	1
Mill automatic power-loom	80	1/16	2	

Source: Sen (1972), appendix C.

THE NATURE OF PRODUCTS

The nature of products in terms of colour, size, cloth, embroidery, produced by the various sectors is so varied that it is almost impossible to generalize. The question may, however, be approached by comparing the nature of products produced by the hand-loom sector in terms of their quality (i.e. in terms of counts of yarn) with those of the mill sector.

The two basic items of dhotis and saris comprise some 52 per cent of total hand-loom products, though these can vary enormously in price depending on the quality. This is shown in Table 2.4 for hand-loom produced dhotis of the same width and length.

The hand-loom sector has an advantage over the mills in the production of the lowest counts since the cost of yarn as a percentage of total cost is highest in these counts. Thus, 'the margin of cost above the raw material is so small that there is relatively less scope for the economies which attend large-scale production'.[14] As the cloth quality improves, weaving costs as a percentage of total costs rise correspondingly with the greater intricacy of the operation so that at the highest counts hand-loom products may be considered skill- as well as labour-intensive.

Competition between the sectors has taken place largely in clothing of medium counts in which mills can take advantage of scale economies,

Table 2.4 Prices and qualities of hand-loom dhotis 1970

Quality of cloth	*Width*	*Length*	*Rupees per unit*
20s/20s	1–27	7–32	11.11
40s/40s	1–27	7–32	12.45
60s/60s	1–27	7–32	15.64
80s/80s	1–27	7–32	19.55
100s/100s	1–27	7–32	21.13

Note: The counts of yarn refer to cloth texture with higher counts representing a finer cloth. The *India Textile Bulletin* classifies counts as follows:

Coarse = Average count yarn less than 17s
Medium B = Average count yarn between 17s and 25s
Medium A = Average count yarn between 26s and 34s
Fine = Average count yarn between 35s and 47s
Superfine = Average count yarn greater than 48s

Source: *India Textile Bulletin*, 1972.

Table 2.5 Percentage distribution of counts of yarn consumed by mills and hand-looms

Counts	*Mills*	*Hand-looms*
1s–10s	8.1	20.0
11s–20s	49.6	34.4
21s–30s	25.7	19.6
31s–40s	11.6	14.2
>40s	5.0	11.8
	100	100

Source: *Report of the Fact Finding Committee*, 1942.

since the proportion of cost above the raw material cost is proportionately higher than that in the lowest counts.[15] Production by mills of the coarse variety is still relatively unimportant as recent figures show.

While it has been found that different counts influence the type of machinery required,[16] there is no precise specification of this relationship. One may, however, postulate that the mills producing the coarser varieties use more dated and thus more labour-intensive processes of production. Some evidence for this is suggested by the fact that the more prosperous mills, which are able to diversify their production, possess the modern finishing and processing equipment necessary for the production of more sophisticated products.[17] The only automatic looms in existence in 1930 produced mainly 'high-class cotton suitings as well made as any in the country'.[18]

RELATIVE PRICES

Data on this question are notoriously scarce. The Textile Enquiry Committee of 1958 noted that 'there is no organization which can throw sufficient and reliable light on what is happening at the consumer's end in regard to the consumption of cloth and price trend'.[19] Indices and prices of mill and hand-loom products are available but are no use for comparing absolute prices, while it is very difficult to make any precise comparison between two products which are differentiated in so many respects. Nevertheless, the broad picture appears to be that while hand-loom products are in general more expensive than the most comparable mill garments,[20] the former are still capable

Table 2.6 Percentage distribution by counts of mill-made clothing

State	*Coarse*	*Medium B*	*Medium A*	*Fine*	*Superfine*	*Total, %*
Andra Pradesh	22.1	45.2	31.4	–	0.5	100
Kerala	6.6	69.3	24.1	–	–	100
Mysore	15.8	39.9	35.8	4.5	4.0	100
Orissa	10.6	28.7	52.2	5.6	2.8	100
Uttar Pradesh	12.5	69.1	17.4	0.1	0.9	100
All India	16.0	32.5	42.2	3.8	4.6	100

Source: Calculated from various issues of the *India Textile Bulletin*.

of producing the very cheapest products made of the coarsest cloth with dull colouring. In a study of cotton clothing in Poona, for example, Brahme found that hand-looms produced the very cheapest saris, purchased by women belonging to the 'Depressed Class'.[21] Another study of Madras found that villagers preferred hand-loom products on the basis of their cheapness and durability.[22]

Ideally, a comparison of relative prices of the products of the two sectors should take place in a situation in which the operative conditions are identical so that any price differential can be attributed to the respective technologies. In reality, however, although hand-loom producers are discriminated against in terms of raw material prices, availability of yarn, interest rates and so on, they receive substantial assistance from the Government. Prasad has estimated the bias in these conditions against hand-loom producers to be some 30 per cent of the sales price.[23] The Fifth Five-Year Plan points out that these difficulties have by no means been removed. On the other hand, the projected outlay on the hand-loom industry in the plan amounts to some 18 per cent of the total value of annual output of the industry. Then again, the controlled cloth scheme has attempted to regulate the price and quantity of certain varieties of mill-made cloth. It is thus extremely difficult to disentangle the purely technological element of prevailing price differentials.

CONSUMPTION OF HAND-LOOM AND MILL CLOTHING BY EXPENDITURE CLASSES

The only source which provides sufficient information to examine the consumption of mill and hand-loom clothing in terms of expenditure classes is the Seventeenth Round of the National Sample Survey, 1961–62.

Except in the highest expenditure classes where woollen and art silk products are purchased, cotton clothing comprises roughly 90 per cent of total expenditure on clothing. In order to establish the labour-intensity of the purchases of the various income and thus expenditure groups, hand-loom, mill-made and khadi-clothing (produced by hand-loom weavers) purchases were summed and each taken as a percentage of the total for rural and urban areas at the all-India level. The results, classified by monthly household expenditure classes are presented in Tables 2.7, 2.8 and 2.9.

Table 2.7 Mill-made clothing as a percentage of the total of mill-made, hand-loom and khadi clothing – classified by monthly household expenditure classes

	0–25	*25–50*	*50–100* *rupees*	*100–150*	*150–300*	*300+*
Rural areas	79.2	82.2	82.9	83.1	88.3	85.8
Urban areas	42.9	84.4	82.4	87.0	83.9	87.0

Source: This table and those that follow have been calculated from the *National Sample Survey* quoted above.

Table 2.8 Hand-loom clothing as a percentage of the total of mill-made, hand-loom and khadi clothing – classified by monthly household expenditure classes

	0–25	*25–50*	*50–100* *rupees*	*100–150*	*150–300*	*300+*
Rural areas	20.8	10.8	15.9	15.7	10.3	12.5
Urban	57.1	15.6	16.2	12.0	2.0	12.0

Table 2.9 Khadi clothing as a percentage of the total of mill-made, hand-loom and khadi clothing – classified by monthly household expenditure classes

	0–25	*25–50*	*50–100* *rupees*	*100–150*	*150–300*	*300+*
Rural areas	–	0	1.2	1.2	1.3	1.1
Urban areas	–	0	1.4	0.8	2.0	1.0

Note: Here and in the following tables dashes represent zero estimates

While no breakdown is available for power-loom products, a study of cotton textile consumption in villages shows the percentage share of the power-loom purchases to be 11 per cent of the total, with hand-loom products and khadi contributing 16 and 3 per cent respectively.[24] The share of hand-looms is slightly higher than the average for rural and urban areas of 13 per cent.

Table 2.8 shows a significantly higher proportion of hand-loom clothing purchased by consumers in the lowest household expenditure class. The exceptionally high figure of 57 per cent in the first urban class reduces to 28 per cent if hand-loom clothing is taken as a percentage of total clothing (if the latter is defined to include bedding and upholstery) since this class spends some 43 per cent of its total clothing expenditure on this item.

It is only consumers well below the average who consume a higher proportion of hand-loom clothing. The average monthly total per capita expenditure in the first household expenditure class is 8.9 rupees for urban areas and 10.1 for rural areas compared to the average for all classes of 30.9 and 21.7 rupees respectively. The percentage of households covered by this first expenditure class is 3 per cent for urban and 4.2 per cent for rural areas.

The proportion of hand-loom clothing purchased drops sharply after the first expenditure class for both the rural and urban sectors. This cross-sectional result should be seen in terms of the historical relationship between products, processes and incomes discussed above. Production by independent weavers of crude products in the earliest period was partially replaced by machinery with the spread of commerce and rise in incomes. But the age structure of machinery in the industry shows that incomes have not risen sufficiently to generate the demand for more sophisticated products and hence processes. The techniques in existence today were designed in advanced countries for the income levels and demand structure appropriate to the early part of this century. We may thus attempt to dynamize the cross-section results above by arguing that as incomes rise, consumers pass, in effect, from the products and processes of the earliest period to those of progressively later periods.[25]

That is to say, they pass from the products and processes of the hand-loom sector to those of the mill sector which become progressively less dated as incomes rise.[26] Technological dualism, then, is not a stark contrast between primitive and advanced products and processes but is rather a continuum of the products and processes of different time periods.

Table 2.10 Proportion of hand-loom clothing to total of mill, hand-loom and khadi clothing – classified by monthly household expenditure classes

	0–25	*25–50*	*50–100 rupees*	*100–150*	*150–300*	*300+*
Madras						
rural	91.7	58.3	55.7	52.2	26.7	23.2
urban	100	50.0	71.4	29.3	31.7	27.7
Mysore						
rural	–	–	5.7	23.4	8.0	35.0
urban	–	–	9.1	16.4	11.7	20.0
Andra Pradesh						
rural	–	16.7	66.7	66.7	28.1	33.6
urban	79.3	0	20.8	20.8	39.0	21.5
Uttar Pradesh						
rural	–	0	4.2	4.2	2.6	2.1
urban	–	–	2.3	2.3	5.2	5.4
Kerala						
rural	100	59.1	38.5	38.5	11.9	22.4
urban	–	67.3	16.4	16.4	26.5	9.3

Finally, the results above at the all-India level conceal tremendous regional variations. The regional variations in the proportion of hand-loom products purchased can be obtained from Table 2.10.

SUMMARY AND CONCLUSIONS

It has been argued that products, processes and incomes are related in each historical period and that the Indian cotton clothing industry reflects a continuum of such relationships. The relationship existing in the pre-capitalist period is reflected in the cross-section data by the mode of production of the hand-loom sector and the higher percentage of its coarse products purchased by the lowest income group. An analysis of the nature of mill products together with the age structure of the machinery suggests the possibility that rising incomes are associated with the product and process characteristics of progressively later periods.

The results have a great deal of relevance for the current concern with income redistribution and employment. Since the bare minimum level of living in India at 1960-61 prices has been defined as 20 and 15 rupees per month[27] for the urban and rural sectors respectively, the results show that an income redistribution aimed at eliminating poverty would, by reducing the hand-loom share of output, substantially reduce employment. This can be seen in terms of an elimination of the lowest household expenditure class in both rural and urban areas. The hand-loom sector is to some extent predicated upon poverty.

This relationship between products, processes and incomes is missed by studies of the employment effects of an income redistribution which not only ignore the self-employed sector but also define products very broadly, i.e. in terms of industries.[28] We have argued that the 'same good' in reality comprises products which are qualitatively different because they are associated with the processes and incomes of different historical periods. It is correspondingly incorrect to speak of 'technological options' producing the 'same good'.[29]

Notes

1. Marx (1946) p. 377.
2. Marx (1964) p. 116.
3. See Stewart (1974).
4. Chenery et al. (1974) p. 12.
5. Self-employment as a percentage of total employment is 59 per cent.
6. Marx (1964) p. 134.
7. Government of India, *Draft Fifth Five-Year Plan 1974–79*, p. 160.
8. Government of India, *Indian Worker – An Industry-Wise Review* (1969).
9. *India Textile Bulletin* (1968).
10. OECD, *Modern Cotton Industry* (1965).
11. It was found in 1942 that of the total of hand-looms, 64 per cent were throw-shuttle and 35 per cent fly-shuttle. See Bagchi (1972) p. 224.
12. *India Textile Bulletin,* Annual Census (1969).
13. According to Bagchi (1972) p. 255, 'The technical backwardness of India is dramatically illustrated by the failure of the mill industry to adopt automatic looms or automatic attachments on any large scale.'
14. Prasad (1963) p. 123.
15. This perhaps explains why the controlled cloth scheme instituted in 1964 has been such a failure. The scheme was designed to ensure that a part of the mill output would be of varieties used by the relatively poor and would be sold at prices they could afford. Instead of encouraging the production of coarse varieties the effect was to divert the output of the mills away from these varieties. See *Economic and Political Weekly,* vol. IX, no. 37 (14 September 1974).
16. National Council of Applied Economic Research, *Demand for Cotton Textile Machinery* (New Delhi, 1967).

17. Kumar et al. (1965).
18. Quoted in Bagchi (1972) p. 255.
19. Government of India, *Report of the Textile Enquiry Committee* (1958), p. 12.
20. Prasad (1963), p. 307, contains a list of prices of comparable mill and hand-loom products drawn from the available evidence.
21. Brahme (1962).
22. 'Consumption pattern study of consumers from Coimbatore District of Madras', reported in *India Textile Bulletin* (1967).
23. Prasad (1963). Similarly, Dantwala in *Economic & Political Weekly*, vol. VII, no. 51 (16 December 1972) argues that 'much of the present non-viability of the production of "inferior" technologies is not entirely due to their inherent inferiority'.
24. Alagh (1974).
25. Cooper's (1973) example of a shift from an ordinary safety razor produced labour-intensively to the consumption of an electric razor produced by highly mechanized methods embodies the same idea.
26. This is not, of course, strictly true since the highest income groups also consume the intricate and luxury hand-loom products. However as Marx pointed out, luxury items are the least important for technological comparison between different economic epochs.
27. Bardhan (1974).
28. The methodology of these studies is summarized in Cline (1975).
29. See Tokman (1975).

References

J. K. Alagh (1974). 'Planning Policy and the Fifth Plan', *Economic and Political Weekly*, Annual no. vol. IX, no. 6, 7 and 8, February.

A. K. Bagchi (1972). *Private Investment in India 1900–1939*, Cambridge University Press.

P. Bardhan (1974). 'On the Incidence of Poverty in Rural India in the Sixties', in T. N. Srinivasan and P. Bardhan (eds), *Poverty and Income Distribution in India*, Indian Statistical Institute.

S. Brahme (1962). *Distribution and Consumption of Cloth in Poona*, Poona.

H. Chenery et al. (1974). *Redistribution with Growth*, Oxford University Press.

W. R. Cline (1975). 'Distribution and Development: a Survey of Literature', *Journal of Development Economics*, February.

C. Cooper (1973). 'Choice of Techniques and Technological Change as Problems in Political Economy', *International Social Science Journal*, no. 3.

D. Kumar, S. P. Nag and L. S. Venkataraman (1965). *Resource Allocation in the Cotton Textile Industry*, Asia Publishing House.

K. Marx (1946). *Capital*, London: Allen & Unwin.

K. Marx (1964). 'Pre-Capitalist Economic Formations' in E. Hobsbawm (ed.), *Pre-Capitalist Economic Formations*, London: Lawrence & Wishart.

K. Prasad (1963). *Technological Choice under Development Planning*, Popular Prakashan.

A. K. Sen (1972). *Choice of Techniques*, Oxford: Blackwell.
F. Stewart (1974). 'Technology and Employment in LDCs', *World Development*, March.
V. Tokman (1975). 'Income Distribution, Technology and Employment in Developing Countries: An Application to Ecuador', *Journal of Development Economics*, March.

APPENDIX TO CHAPTER 2
FURTHER EVIDENCE CONCERNING THE RELATIONSHIP BETWEEN PRODUCTS, PROCESS AND INCOMES

Chapters 2 and 3 contain evidence of the relationship between products, processes and incomes that conforms closely to what is predicted by our analytical framework. Especially since one can point to counter-examples such as footwear, where the products of relatively capital-intensive technologies are sometimes consumed by low, rather than high-income households, these two cases alone however are unlikely to convince the sceptical reader. The purpose of this appendix, accordingly, is to provide further evidence of the postulated relationship between products, processes and incomes. Let us begin with the evidence relating to India, since it is that country that is the subject of Chapters 2 and 3.

Manufactures in India

Whereas the two chapters are concerned respectively with textiles and sweeteners, other manufactured goods in India appear to exhibit similar patterns. In the case of cigarettes and milk, for example, it has been suggested that 'much of rural consumption is in a less-processed form, supplied by small-scale producers utilizing less capital-intensive techniques than those associated with the higher income, large-scale urban markets'.[1] In the market for laundry soap, as well,

> Variations in the quality ... over a wide range within the non-power sector and between the non-power and power sectors, and sizable price differentials point to a highly segmented product market. Interviews with sample firms during the 1982 survey revealed that the socio-economic status of the consumers of laundry soap varied directly with the quality and the price of the product. Low varieties

of soap manufactured by the tiny and medium size manually operated firms were consumed mostly by poorer households (including industrial workers, coolies and other unskilled workers), while medium varieties... produced by large non-power firms or power sector firms were sold to upper-middle and high-income households. The power sector firms, in particular, developed their markets among the affluent sections characterised by low price elasticity of demand.[2]

The soap industry in Bangladesh and Ghana

Mubin and Forsyth have conducted a detailed study of the relationship between the production and consumption characteristics of different types of soap in Bangladesh and Ghana.[3] In the former country, three distinct types of soap were identified, namely, toilet soap that is produced only in capital-intensive plants and washing soap that is manufactured by both labour- and capital-intensive methods.

Among other respects, the labour-intensive techniques differ from the capital-intensive alternatives with regard to their degree of dependency on imported inputs. For whereas the former require 'many local inputs, of raw materials and from engineering firms',[4] the latter depend 'heavily on imported raw materials, plant and equipment'. 'Thus' conclude the authors, 'if we consider the spread effects of the choice of technology on the Bangladesh economy, these are likely to be much greater in the case of the labour-intensive technology.'[5] (This aspect of the difference between traditional and modern technological systems will be dealt with later in the chapters dealing with the macroeconomic modelling of these systems).

A second major distinction between the alternative methods of soap production emerges from Table 2A.1 which shows that the soaps produced by labour- and capital-intensive techniques are favoured by systematically different categories of households.

Toilet soap, for example, possesses a comparative abundance of high-income characteristics such as packaging and scent and thus tends to be purchased primarily by upper-income households. In the case of washing soap, 'it is apparent that the shift to capital-intensive washing soaps is related to income levels, with the balance of consumption shifting heavily in [its] favour as income rises (especially in the urban areas).'[6] 'Preferences for soap', conclude Mubin and Forsyth, 'are not random. On the contrary, there is a clear tendency for low incomes, rural location and manual

Table 2A.1 Distribution of consumption by soap type and income class (%)

Household income (Tk/month)	*Washing soap (labour-intensive)*			*Washing soap (capital-intensive)*			*Toilet soap (capital-intensive)*			*Total soap*[*]			*Proportion of total consumption for the country as a whole*	
	Rural	*Urban*	*Total*	*Rural*	*Urban*	*Total*	*Rural*	*Urban*	*Total*	*Rural*	*Urban*	*Total*	*By group*	*Cumulative*
0–99	–	–	–	–	–	–	–	–	–	–	–	–	–	–
100–149	100.0	–	100.0	–	–	100.0	–	–	100.0	100.0	–	100.0	0.5	0.5
150–199	100.0	–	100.0	–	–	100.0	–	–	100.0	100.0	–	100.0	3.0	3.5
200–249	76.8	–	76.8	5.9	–	5.9	17.2	–	17.2	100.0	–	100.0	4.8	8.3
250–299	69.2	–	69.2	30.8	–	30.8	–	–	–	100.0	–	100.0	5.8	14.1
300–399	78.7	0.0	74.1	19.2	100.0	24.0	2.1	–	1.9	100.0	100.0	100.0	12.9	27.0
400–499	61.1	11.7	57.6	26.6	62.5	29.1	12.3	25.8	13.3	100.0	100.0	100.0	11.7	38.1
500–749	48.3	23.5	46.4	43.7	48.6	45.1	8.0	27.8	9.5	100.0	100.0	100.0	21.8	60.5
750–999	25.3	7.1	23.7	66.2	61.4	65.7	8.5	31.4	10.5	100.0	100.0	100.0	12.6	73.1
1000–1499	52.7	2.6	46.3	7.0	60.4	13.9	40.3	36.8	39.0	100.0	100.0	100.0	17.1	90.2
1500–1999	69.0	3.6	61.1	18.6	54.1	22.9	12.4	42.3	16.0	100.0	100.0	100.0	4.9	95.1
2000 and over	12.6	–	10.9	50.3	62.9	52.0	37.2	37.1	37.1	100.0	100.0	100.0	4.8	99.9
Total			52.4			32.0			15.6			100.0		100.0[*]

[*]Error due to rounding.

Source: Mubin and Forsyth (1984).

occupations to be associated with the purchase of labour-intensive soap; and for high incomes, urban location, and non-manual occupations to be associated with the purchase of capital-intensive soaps.'[7]

Very similar patterns emerged in the case of Ghana, where the 'same basic soap types identified as important in Bangladesh were again central... labour-intensive washing soap was purchased principally by the lower-income earners, and as household incomes rose, so consumption of capital-intensive washing soap and capital-intensive toilet soap also rose'.[8] And again as in Bangladesh, the varying factor intensities of production were associated with the numerous other features of technological dualism. In particular, 'a modern, large-scale, capital-intensive sector, drawing most of its equipment from abroad' co-existed with 'a locally oriented, small-scale, labour intensive sector'.[9]

Roof-tiles in Indonesia

In Central Java the production of roof-tiles takes place on the basis of traditional labour-intensive techniques as well as relatively large-scale, mechanized methods. A recent study of these alternatives by Sandee is especially useful for our purposes because it shows that they are 'embedded' in distinct systems of production and consumption.[10] Indeed, as summarized in Table 2A.2, Sandee's findings bear a distinct similarity to those reported for the textile industry in Chapter 2. For what can clearly be seen from that table is that the four roof-tile techniques under consideration form part of a continuum along each of the rows, representing some of the key dimensions of a technological system.

At the one extreme, one finds open-fire technology, which forms part of a distinctly pre-capitalist mode of production, based on family labour and a limited need for intermediaries 'since producers can easily deal with consumers directly'.[11] Those consumers are generally low-income rural households. At the other extreme, the most mechanized technique (the power press) is based on wage labour in large firms, selling to middle and high-income consumers in urban areas. The two other techniques occupy intermediate positions along the rows of the table, thus emphasizing the historical continuity in the evolution of roof-tile technology.

Table 2A.2 Alternative roof-tile technologies and technological systems

	Open-fire	*Kiln*	*Hand press*	*Power press*
Gross value added per unit (rupees)	7.1	9.1	23.5	45.0
Labour	Family	Family and wage workers	Family and wage workers	Wage workers
Scale of enterprise	Household	Micro enterprise	Small	Medium
Marketing	Producers, Small traders	Producers, Middlemen	Middlemen, Agents	Agents
Markets	Poor rural consumers	Rural and poor urban consumers	Middle income urban consumers	Middle income and high income urban consumers
Relative importance in subsector employment generation	75%		12%	10%

Source: Sandee (1995).

The history of industrialization in Nigeria

Kilby's account of the process of industrialization in Nigeria over the period from 1945 to 1966 contains several examples of how traditional and modern technological systems coexisted in the industrial sector.[12] Kilby discusses these under the heading 'market segmentation', from which the following quotation is extracted.

> For a great many products there is not a single demand, but a number of demands for varying qualities of the item at different prices. Such 'product differentiation' is to be found in the markets of developed countries but the range in quality and price is comparatively narrow, the difference in price seldom exceeding, say, 100%. In underdeveloped countries this range is very much greater. To give a few examples, an arm-chair made in a small-scale Nigerian workshop sold in 1964 for £2 while the same model chair – but constructed from seasoned wood, squarely joined and perfectly flush on the floor – sold for over £6 from any one of the country's major furniture manufacturers. In the case of sandals there is a wide range of possibilities starting with the tyre-sandal at 2*s* and 6*d* (cut from

discarded automobile tyres with a plastic thong tacked on) and going up to the 55*s*. English 'cross-sandal'. The soap market is supplied by black wood-ash soap from cottage producers, cold process caustic soda soap from modern small-scale Nigerian firms at a price about 80% above that of black soap, and finally hot process emulsion soap from large-scale expatriate-run factories at double the price of the cold process soap.[13]

Furniture-making in Kenya

Using a considerable amount of detailed survey data for Kenya, House has tried to test the following hypothesis concerning the relationship between products, processes and incomes: 'Do low-income households show a greater tendency to buy from the labour-intensive informal sector, and is it generally the case that higher-income households buy superior-quality furniture from the less labour-intensive formal sector?'[14]

What he found was no less clear a pattern than that associated with the other evidence described in this appendix, namely, that:

> The poor households in Nairobi, and by inference in other urban areas and in rural Kenya also, buy low-quality, relatively inexpensive furniture from the very labour-intensive informal sector. As incomes rise, preferences move in favour of more expensive and middle-quality furniture, produced in the small-scale workshops and large-size factories. Both kinds of producers are less labour-intensive than informal-sector producers but they are still very labour-intensive judged against other branches of industry. And both use a very small amount of imported raw materials.
>
> At the top of the income scale, among the prosperous business and professional elite, are found the buyers of the luxury items of furniture. These are the very expensive, highly skill and import-intensive products which have been identified with the firms in the 50–99 size-class, which exhibit well above average levels of labour productivity.[15]

Notes to Appendix

1. Johnston and Kilby (1975) p. 307.
2. Suri (1988) p. 217.
3. Mubin and Forsyth (1984).
4. Ibid., p. 24.
5. Ibid.
6. Ibid., p. 39.
7. Ibid., p. 44.
8. Ibid., p. 51.
9. Ibid., p. 50.
10. Sandee (1995).
11. Ibid., p. 61.
12. Kilby (1969).
13. Ibid., p. 33.
14. House (1984) p. 156.
15. Ibid., p. 158.

References

W. House (1984). 'Product Choice and Employment in Furniture-Making in Kenya', in W. van Ginneken and C. Baron (eds), *Appropriate Products, Employment and Technology*, London: Macmillan.

B. Johnston and P. Kilby (1975). *Agriculture and Structural Transformation*, Ithaca: Cornell University Press.

P. Kilby (1969). *Industrialization in an Open Economy: Nigeria 1945–1966*, Cambridge University Press.

A. Mubin and D. Forsyth (1984). 'Technology, Employment and Income Distribution: the Soap Industry in Bangladesh', in W. van Ginneken and C. Baron (eds) *Appropriate Products, Employment and Technology*, London: Macmillan.

H. Sandee (1995). *Innovation Adoption in Rural Industry: Technological Change in Roof Tile Clusters in Central Java, Indonesia.* PhD dissertation, Free University, Amsterdam.

K. B. Suri (1988). 'Technology, Firm Size and Product Quality: a Study of Laundry Soap in India', in K. B. Suri (ed.) *Small-Scale Enterprises in Industrial Development: the Indian Experience*, London: Sage.

3 The Employment Effects of an Income Redistribution: a Test for Aggregation Bias in Indian Manufacturing

INTRODUCTION

The ILO Mission Report on Colombia espoused the idea that since the poor tend to consume relatively more labour-intensive commodities than the rich, a redistribution in their favour will tend to increase employment.[1] This proposal led subsequently to a host of attempts to demonstrate its empirical validity at the economy-wide level in a number of less-developed countries.[2] The results, however, gave only qualified support to the hypothesis. As Morawetz put it, 'The almost unanimous conclusion is that even quite significant redistributions of income seem likely to have only marginal effects on growth and employment, usually increasing the latter by less than 5%.'[3]

This chapter is concerned to examine one possible reason for these disappointing results, namely the aggregation of products into broad, usually two-digit categories. Aggregation in this sense involves an estimation bias if there are shifts *within* these categories to goods produced by different factor combinations. In this study we attempt to examine the nature and extent of aggregation bias within a particular industry – that of sugar processing in India.

PRODUCTS AND TECHNOLOGIES IN THE INDUSTRY

Sugar processing in India is a good example of an industry in which two similar but distinct products serve the different ends of the market. A traditional product, gur, which is basically sugar cane in concentrated form is still the dominant sweetener for the vast majority of those with low incomes, particularly in rural areas. Some 60 per cent of the annual sugar cane crop is used for gur manufacture and 28 per cent

* First published by the International Labour Office in the World Employment Programme Research Working Paper Series, November 1977.

of the remainder goes to the refined sugar industry. Annual per capita consumption of gur is roughly 12 kilos as opposed to 5 kilos in the case of sugar.[4] Gur represents on average 1.5 per cent of total consumption expenditure in rural areas and 0.6 per cent in urban areas. The consumption patterns in the industry can be represented as in Table 3.1 which shows the data presented by monthly per capita expenditure classes for rural and urban areas in terms of index numbers where average expenditure for all classes on gur and sugar is set equal to 100 and the expenditure in each class expressed as a percentage thereof.[5]

The manufacture of gur can be divided into the three stages of extraction, clarification and evaporation of the cane juice.[6] The earliest method of crushing consisted of grinding the cane with a pestle and mortar with the pestle moved by bullocks. This method was followed by crushing the cane between two wooden or iron rollers fitted close to one another vertically or horizontally. Such crushers driven by bullocks are still typically employed by small-scale producers. Power-crushers of varying designs and sizes are used by the small number of large-scale manufacturers.[7] Following the process of extraction the juice is clarified and then concentrated by boiling and evaporation. The resultant semi-liquid mass, when cooled, solidifies into gur.

The production of white sugar differs in that it involves the centrifugal separation of sugar crystals from molasses.[8] Crystal sugar is produced by both vacuum pan and open pan sulfitation techniques, the latter being associated with smaller-scale units than the sugar mills proper. The proportion of total output accounted for by open-pan producers is very small (roughly 5 per cent).[9]

Data on labour and capital per unit of output for the two products are taken from Prasad as shown in Table 3.2. Gur thus uses more labour but less capital in relation to output than sugar. The latter is roughly double the price of the former as Table 3.3 shows for rural and urban areas. When the figures for labour productivity in Table 3.2 in terms of tons are adjusted to reflect relative prices, the higher sugar price raises labour productivity in sugar as opposed to gur manufacture. Since the ratio of value-added to gross output is also higher for sugar (25.2 per cent) than gur manufacture (13.4 per cent),[10] labour productivity measured in terms of labour per unit of value added would further increase the differential productivities. With capital productivity the adjustments of course work in the opposite direction though capital per unit value added in gur production (17.7) remains below that for sugar (19.5).[11]

Table 3.1 Consumption by per capita expenditure classes

	Monthly per capita expenditure classes (rupees)													
	0–8	*8–11*	*11–13*	*13–15*	*15–18*	*18–21*	*21–24*	*24–28*	*28–34*	*34–43*	*43–55*	*55–75*	*75+*	*Average*
Rural (all India)														
Numbers in each class as per cent	4.39	11.36	9.51	10.77	14.6	12.57	9.2	8.19	7.31	5.82	3.23	1.72	1.33	
Gur	15.6	40.6	49.9	62.5	87.5	90.6	103.1	121.9	153.1	206.3	221.9	340.6	350.0	100
Sugar	3.8	15.4	30.8	38.5	57.7	69.2	96.2	138.5	169.2	238.5	269.2	403.8	980.8	100
Urban (all India)														
Numbers in each class as per cent	2.28	4.92	5.14	7.28	10.34	11.39	10.31	9.44	11.63	9.41	7.75	5.18	4.93	
Gur	22.2	61.1	66.7	88.9	77.8	111.1	88.9	105.6	105.6	111.1	172.2	122.2	100	100
Sugar	10.1	17.4	30.4	40.6	53.6	71.0	79.7	94.2	114.5	146.4	168.1	202.9	249.3	100

Source: Indian National Sample Survey (1961–2).

Table 3.2 Labour productivity, gur and sugar

	Gur	*Sugar*
Output per person in 100 days	2.1 tons	8 tons
Total productive capital per ton	236.79 rupees	982.00 rupees
Total productive capital per person	503.25 rupees	7859.00 rupees

Source: Prasad (1963) pp. 99–100.

Table 3.3. Prices, gur and sugar

	Rural	*Urban*
Gur*	0.48	0.55
Sugar*	1.13	1.06

**Note:* Rupees per seer. A seer is approximately equal to one kilogram.

Source: Calculated from Indian National Sample Survey.

THE SIMULATION METHODOLOGY

In the subsequent sections an attempt is made to estimate the employment effects of alternative simulated income redistributions using both aggregative and disaggregative approaches. Before beginning the estimation procedure, however, it is useful to set out the broad simulation methodology which underlies this study. This can be expressed as follows.[12] Given a number of total per capita expenditure classes n, average per capita expenditure in each class $\bar{Y}_i (i = 1, \cdots, n)$ and the percentage of population in each class $\bar{N}_i (i = 1, \cdots, n)$ and $\sum \bar{N}_i = 100$) total expenditure can be expressed as:[13]

$$Y = \sum_{i=1}^{n} \bar{Y}_i \bar{N}_i, \tag{3.1}$$

average per capita expenditure as:

$$Y_w = \sum_{i=1}^{n} \frac{\bar{Y}_i \bar{N}_i}{N}, \tag{3.2}$$

where $N = 100$. In the case of equal numbers in each group,

$$Y_w = \sum_{i=1}^{n} \bar{Y}_i | n,$$

otherwise as here,

$$Y_w \neq \sum_{i=1}^{n} \bar{Y}_i | n \tag{3.3}$$

The share of each group in total expenditure:

$$K_i = \frac{\bar{Y}_i \bar{N}_i}{Y}, i = 1, \cdots, n \tag{3.4}$$

The simulation takes the form of changing the elements of the vector comprising the shares K_i with Y and $\bar{N}_i$ given.[14] As can be seen from (3.4) the result is a new $\bar{Y}$ for each class, i.e. a new level of average per capita expenditure in each class. An estimated consumption function for each commodity concerned is applied to these new average per capita expenditure levels for each class. The summation across the classes yields a new estimated total consumption expenditure on the commodity which may be compared with the pre-redistribution total to arrive at the percentage change.

ESTIMATES OF CONSUMPTION FUNCTIONS

In choosing the best-fitting function applied to the thirteen expenditure classes in Table 3.1 one problem is that there are unequal numbers of persons in the different classes. This presents a problem in estimation inasmuch as the sum of the residuals weighted by these numbers is not zero (though the unweighted residuals do sum to zero).

In order to get the sum of the weighted residuals to sum to zero some form of weighting is required. This was achieved by weighting each of the initial thirteen observations by the numbers in each class, i.e. effectively raising the total number of observations in the proportions $\bar{N}_i (i = 1, \cdots, n)$. The functions estimated on the basis of the weighted regression analysis are shown below:

$$\text{Gur (rural) } Y = \underset{(-17.62)}{-0.0001x^2} + \underset{(36.02)}{0.023x} \underset{(-9.05)}{- 0.095}, \quad \bar{R}^2 = 0.976$$

Sugar (rural) $Y = 0.000056x^2 + 0.016x - 0.12, \qquad \bar{R}^2 = 0.994$

$$(11.56) \quad (30.15) \quad (-13.83)$$

Gur (urban)[15] (1) $\log Y = -1.32 - \frac{9.18}{x}, \bar{R}^2 = 0.79$

$$(-31.17)(-12.6)$$

(2) $Y = -0.000042x^2 + 0.0059x + 0.063, \quad \bar{R}^2 = 0.682$

$$(-8.08) \quad (9.24) \quad (4.91)$$

Sugar (urban) $Y = -0.00018x^2 + 0.037x - 0.18, \qquad \bar{R}^2 = 0.995$

$$(34.06) \quad (58.46) \quad (-14.14)$$

(t values in parentheses)

Since the functions are strictly concave (the coefficient of x^2 is negative) an income redistribution towards the average will in each case increase total consumption except for sugar in rural areas where there will be a decrease.[16] In the case of the alternative log-reciprocal function for urban gur nothing a priori can be said.

THE CONSUMPTION AND EMPLOYMENT EFFECTS OF AN INCOME REDISTRIBUTION

A series of income redistributions have been simulated by changing the shares of total expenditure accruing to the thirteen expenditure classes.[17] Tables 3.4 and 3.5 show the new configurations of income between classes separately for rural and urban areas (with total income given and constant). For each simulation the associated Lorenz ratio is shown at the foot of each table, together with the percentage of total income transferred.

Simulations (1), (2) and (3) are designed to eliminate all those below the absolute poverty line of 15 rupees per month[18] (i.e. the bottom four classes are eliminated), and simulation (7) represents complete equality in the sense of a Lorenz ratio equal to zero.

The tables show that the elimination of absolute poverty in this period would have required a transfer of income representing 10 per cent of the total income in urban areas and 6.2 per cent in rural.[19] The implied reduction in the Lorenz ratio is 32 per cent (rural) and 39 per cent (urban). There is little reason to suppose that these figures would

Table 3.4 Rural simulation

Expenditure class	*Original share in total expenditure as percentage*	*Simulation*						
		(1)	*(2)*	*(3)*	*(4)*	*(5)*	*(6)*	*(7)*
0–8	1.3	3.0	3.0	3.0	2.0	2.0	3.0	4.4
8–11	5.0	7.8	7.8	7.8	6.0	5.5	5.0	11.4
11–13	5.3	6.6	6.6	6.6	6.0	5.5	5.3	9.5
13–15	6.9	7.4	7.4	7.4	7.0	7.0	6.9	10.8
15–18	11.0	11.0	10.2	11.0	12.0	11.0	11.0	14.6
18–21	11.3	11.3	10.4	11.3	12.0	11.3	11.3	12.6
21–24	9.5	9.5	8.8	9.5	11.0	9.5	9.5	9.2
24–28	9.8	9.0	9.0	9.8	10.0	9.8	9.8	8.2
28–34	10.3	9.5	9.5	9.1	10.0	10.3	10.3	7.3
34–43	10.1	9.1	9.3	8.9	8.0	10.1	10.1	5.8
43–55	7.2	6.2	6.6	5.9	6.0	7.0	7.2	3.2
55–75	5.0	3.7	4.6	3.7	4.0	4.5	5.0	1.7
75+	7.2	5.9	6.8	6.0	6.0	6.5	5.6	1.3
	100	100	100	100	100	100	100	100
Lorenz ratio	0.31	0.21	0.23	0.21	0.25	0.29	0.28	0.0
Percentage of total income transferred		6.2	6.2	6.2	5.9	1.4	1.7	22.4

be much altered at the present time if the poverty line were adjusted upwards by an index of prices relevant to the absolute poor.[20]

The consumption effect – disaggregated basis

The percentage changes in consumption for gur and sugar can be estimated as follows: the changed shares in total income of Tables 3.4 and 3.5 imply new average total per capita incomes in each class and the chosen consumption functions are applied to these for each class. Multiplying the outcome by the given (and constant) numbers in each class and summing over all classes yields the estimated new total demands. These are shown in Table 3.6. The last column in Table 3.6 represents the estimates for gur using the alternative log-reciprocal function. They do not differ much from those based on the quadratic function.

Table 3.5 Urban simulation

Expenditure class	*Original share in total expenditure as percentage*	*Simulation*						
		(1)	*(2)*	*(3)*	*(4)*	*(5)*	*(6)*	*(7)*
0–8	0.4	1.7	1.7	1.7	1.2	2.0	2.0	2.3
8–11	1.6	3.6	3.6	3.6	2.5	2.0	1.6	4.9
11–13	2.0	3.7	3.7	3.7	3.0	2.0	2.0	5.1
13–15	3.3	5.3	5.3	5.3	3.9	3.3	3.3	7.3
15–18	5.5	7.5	7.5	7.5	6.0	5.5	5.5	10.3
18–21	7.2	8.3	8.3	8.3	8.0	7.2	7.2	11.4
21–24	7.5	7.5	7.5	7.5	7.9	7.5	7.5	10.3
24–28	7.9	7.9	6.8	7.9	8.2	7.9	7.9	9.4
28–34	11.6	11.6	10.0	9.6	11.8	11.6	11.6	11.6
34–43	11.5	9.9	9.9	9.5	10.0	11.5	11.5	9.4
43–55	12.2	10.4	10.5	10.2	11.5	12.2	12.2	7.7
55–75	10.7	8.6	9.8	8.7	10.0	10.0	10.7	5.2
75+	18.6	14.0	16.0	16.5	16.0	17.3	17.0	4.9
	100	100	100	100	100	100	100	100
Lorenz ratio	0.36	0.2	0.22	0.22	0.28	0.32	0.33	0.0
Percentage of total income transferred		10	10	10	5.6	2.1	1.6	25.8

The table shows an increased demand for gur in both cases which is significantly high for urban areas. The predicted decline in consumption of sugar in rural areas as opposed to its increase for the urban sector reflects the much higher initial consumption in the latter case and the marked shift away from gur in the middle and upper income classes in the rural case. The strict convexity of the function in the case of rural areas (i.e. the rising mpc) is dependent upon the extremely high estimate for the last class (see Table 3.1). Since this figure for the combined sample estimate is based on widely divergent sub-samples[21] and in view of the fact that without this observation the linear (convex) function gives a fractionally better fit[22] it is likely that the predicted decline in consumption is an overstatement.

By taking a weighted average of rural and urban areas the composite total consumption effect can be obtained. The weights used are the shares of rural and urban sectors in the total output of gur and sugar in

Table 3.6 Change in demand

Simulation	*Percentage change in demand for gur* Rural		Urban		*Percentage change in demand for sugar* Rural		Urban		*Estimates for gur*
(1)	+3.6	(0.22)	+7.9	(0.07)	-2.4	(0.35)	+8.6	(0.18)	+7.8
(2)	+2.0	(0.24)	+5.7	(0.06)	-1.3	(0.39)	+6.3	(0.18)	+6.8
(3)	+3.6	(0.22)	+5.5	(0.06)	-2.4	(0.35)	+5.7	(0.18)	+7.0
(4)	+3.1	(0.27)	+4.6	(0.10)	-2.1	(0.40)	+5.0	(0.26)	+4.9
(5)	+1.5	(0.31)	+2.3	(0.11)	-1.0	(0.47)	+2.5	(0.29)	+2.4
(6)	+2.1	(0.30)	+2.2	(0.12)	-1.4	(0.46)	+2.4	(0.30)	+1.9
(7)	+7.7	(0)	+13.2	(0)	-5.2	(0)	+14.6	(0)	+11.8

Notes: Pre-redistribution specific concentration ratios are 0.51 for sugar and 0.33 for gur in rural areas. In urban districts the figures are 0.33 and 0.12 for sugar and gur respectively. Figures in brackets denote the specific concentration ratios associated with each simulation. They are simply the Lorenz ratios for particular commodities.

Table 3.7 Weighted change in demand

Simulation	*Percentage change in demand for gur*	*Percentage change in demand for sugar*
(1)	+5.0	+5.7
(2)	+3.2	+4.3
(3)	+4.2	+3.6
(7)	+9.5	+9.5

1961–2.[23] These combined results are presented in Table 3.7 for the absolute poverty simulations (1), (2) and (3) and the totally egalitarian case as these are directly comparable.

The fall in the demand for sugar in rural areas is thus offset by the high share of urban areas while the large rural share in gur consumption pulls down the higher predicted increase in urban consumption.

An alternative set of calculations was performed on the absolute poverty simulations for gur using the discrete data instead of the continuous function and making the *ex ante* mpc assumption.[24] In both rural and urban areas the results were very similar to those shown in Table 3.6.[25] The main difference between the methods lies not in the overall estimates but in the fact that with a continuous (and strictly concave) function any redistribution which is an unambiguous movement towards the average increases consumption while in the discrete case this does not always hold because of the discontinuities in the behaviour of the marginal function.

The employment effect

With the assumption of a constant labour–output ratio, employment increases at the same percentage rate as output. To define the increase in terms of actual numbers, however, requires applying the labour–output ratios for gur and sugar to the estimated incremental output. Taking the output of gur and sugar in 1961–2, calculating the rural and urban shares and applying the percentage increases in demand of the last section yields the incremental output. Multiplying by the labour-output ratios of 7:100 tons for sugar and 47:100[26] tons for gur gives the additional numbers demanded for each simulation. Table 3.8 presents these results for rural and urban areas.

Table 3.8 Change in demand for labour

Change in demand for labour						
	Gur		*Sugar*		*Gur*	*Sugar*
Simulation	*Urban*	*Rural*	*Urban*	*Rural*	*Urban+rural*	*Urban+rural*
(1)	(+71,205) +72,118	+66,724	+11,473	–1,112	+138,842	+10,361
(2)	(+62,076) +52,034	+37,068	+8,308	–602	+89,102	+7,706
(3)	(+63,902) +50,208	+66,724	+7,517	–1,112	+116,932	+6,405
(4)	(+44,731) +41,992	+57,456	+6,594	–973	+99,448	+5,621
(5)	(+21,909) +20,996	+27,801	+3,297	–463	+48,797	+2,834
(6)	(+17,344) +20,083	+38,922	+3,165	–648	+59,005	+2,517
(7)	(+107,721) +120,501	+142,715	+19,122	–2,409	+263,216	+16,845

**Notes*: Actual numbers. Figures in brackets represent estimates based on the log-reciprocal function for urban gur.

THE AGGREGATIVE APPROACH – THE DIRECTION AND EXTENT OF AGGREGATION BIAS

Unlike the disaggregative approach used above which estimates the change in consumption and employment for gur and sugar separately, the aggregative approach groups the two products together into a single category of sweetener and applies a consumption function estimated for the two products combined.[27] We take first the situation in the rural areas. To ease the computational requirements it was decided to apply the consumption function for the composite good to the changed income levels associated with just one simulation – that of perfect equality with a Lorenz ratio equal to zero. On this basis, consumption of the composite good was found to increase by 3.4 per cent. Since this is equivalent to growth of both gur and sugar at 3.4 per cent the incremental outputs of the two were obtained in the manner described for the disaggregative approach. Labour and capital requirements were then found using the same labour/output and capital/output ratios as in the disaggregated approach. Table 3.9 shows the differential increases in capital and labour predicted by the two approaches on the basis of simulation (7).

The aggregative approach thus understates both capital and labour requirements relative to the approach which considers the shifts between the two products. In the case of labour the understatement is extremely marked.

In the urban sector it is plain from Table 3.6 that there is an almost equi-proportionate rise in the two products so that no aggregation bias results. This result can probably be explained by the fact that sugar typically has a satiety level since there is no (or very little) shift to higher qualities as incomes rise.[28] In their extensive study of English family budgets Prais and Houthakker thus found that of all the commodities studied, only in the case of sugar were no such shifts apparent.[29]

Table 3.9 Aggregation bias

	Disaggregative approach	*Aggregative approach*	*Aggregation bias*
Increase in labour (actual numbers)	140,306	64,592	54%
Increase in capital (millions of rupees)	53.0	44.1	17%

CONCLUSIONS

This chapter has examined the problem of aggregation bias in the use of existing models which purport to estimate the employment effects of an income redistribution. That is, it has examined for one industry the distortions resulting from the aggregation into a homogeneous group products which in fact are produced by highly disparate methods of production and consumed by different income groups. In the Indian sugar processing industry it was found that while no bias exists for the urban sector, aggregative estimates for the rural areas involve significant distortions. In particular it was found that labour requirements were considerably understated relative to the results reached by the disaggregative approach. Understatement was also found in the case of capital albeit to a lesser extent. Obviously only very limited conclusions can be drawn on the basis of results from a single industry in a single country. However, similar relationships found between different types of products, technologies and incomes in other industries suggest that our results may have wider significance.[30] In any event there is a clear need for further studies on the basis of disaggregated data to test the ubiquity and direction of aggregation bias for various industries and countries.

Notes

1. ILO (1970).
2. These are surveyed in Cline (1975).
3. Morawetz (1974) p. 506.
4. *Indian Sugar Yearbook* (1960–70) p. 119.
5. A method of presentation developed by Mahalanobis (1958). In all cases here and in what follows estimates used are those of the combined sample in the Indian National Sample Survey (1961–2).
6. Gur manufacture is described in Government of India (1965) and Roy (1951).
7. Of the total number of sugar cane crushers in use 96 per cent are worked by bullocks. Calculated from Government of India (1959–60).
8. The techniques of production of white sugar are described in detail in Baron (1975).
9. Baron (1975).
10. Ratios of value-added to gross output were calculated from data given in Prasad (1963).
11. Assuming a price differential of 2:1 between sugar and gur.
12. A similar formulation for equal numbers in each class is to be found in Paukert, Skolka and Maton (1976).
13. Instead of as here regarding the total population as 100, actual numbers can be used so that actual total expenditure is $\sum_{i=1}^{n} \bar{Y}_i \bar{N}_i K$, where K is a scalar.
14. An alternative is to regard $\bar{N}_i$ as variable so that an increase in income for a group is reflected in fewer numbers in the group. Meade characterizes this as 'contraction of the poor' as opposed to 'enrichment'. (See Meade (1976).) Where, however, the distribution of individual house-

holds about the mean income is unknown for each group this procedure is difficult.

15. In view of the relatively inferior fit in the case of urban gur two alternative functions have been presented (and used in the simulations). Strictly speaking the adjusted coefficients of determination ($\bar{R}^2$) of these two functions cannot be compared since the dependent variables (Y and log Y) are different. Goldberger suggests an adjustment procedure to put the functions on the same footing so that $\bar{R}_2$ can be used as the choice criterion in choosing the best fitting function. See Goldberger (1964), p. 217.
16. Strict concavity of a consumption function means that linear interpolation between any two points always gives lower values for consumption than the function itself. An analogous use is made of strict concavity by Sen in the context of utilitarian welfare economics. See Sen (1973).
17. The simulations are based on the assumption that the income redistributions do not alter the price relative between sugar and gur which is given in estimating the functional forms.
18. This is defined at 1960–1 all-India prices as 15 and 20 rupees per month for rural and urban areas respectively. See Bardhan (1974). Dandekar and Rath use a slightly higher figure of 22.5 rupees for urban areas. See Dandekar and Rath (1971). The figure of 21 rupees used here may be regarded as an average of these estimates.
19. Dandekar and Rath regard this figure in absolute terms as a conceptual measure of unemployment and under-employment inasmuch as it represents the amount of additional employment needed to remove absolute poverty. For 1968–9 the figure for rural areas is 822.4 rupees crore. See Dandekar and Rath (1974).
20. Over the period 1958–68 it has been suggested that 'disparities in consumption in real terms increased'. See Chatterjee and Bhattacharya (1974). Bardhan finds that 'between 1960–1961 and 1967–1968 there has been a very significant rise in the percentage of people below the minimum level'. See Bardhan (1974), p. 276.
21. The combined sample estimate is 2.55 rupees and the three subsample estimates are 1.18, 1.62 and 3.89 rupees.
22. Though this is also true of gur in rural areas the sub-sample estimates are far more convergent and hence more reliable. The combined estimate is 1.12 rupees, the average of 1.15, 1.05 and 1.16 rupees.
23. Total production of gur in 1961–2 was 5,885,827 tons, 67 per cent of which was demanded by rural areas. In the same period output of sugar comprised 2,546,000 tons; the urban sector dominating (74 per cent) the consumption in this case. Total production estimates are from the *Indian Sugar Yearbook*, 1969–70. The rural and urban shares are taken from the average consumption of gur and sugar in volume terms presented in the Indian National Sample Survey 1961–2.
24. The *ex ante* mpc assumption is that for each individual the change in consumption is equal to the change in income multiplied by the pre-redistribution mpc. The difficulty with this assumption is that it makes ΔC insensitive to the size of ΔY. We have used in the simulations the *ex post* apc assumption, i.e. each individual in moving to a new level

assumes the apc of those already at that level. With this assumption the mpc is not insensitive to the size of ΔY. Our approach follows Cline (1972).

25. For rural gur the percentage increases in consumption were 3.2, 2.4 and 3.0 per cent for simulations (1), (2) and (3) respectively. For urban gur the corresponding estimates were 8.3, 7.6 and 8.3 per cent.
26. The labour–output ratio for sugar shown in Table 3.7, which is based on a 100-day season, has here been reduced to reflect the fact that the season varies between 135 and 200 days. (See Baron (1975).) Gur, on the other hand, operates on the basis of a 100-day season. The capital–output ratio for sugar has also been reduced in the same proportion to a figure of 549.92 from 982 rupees.
27. The estimated consumption function for the rural areas is

$$y = -\underset{(-5.7)}{0.000046x^2} + \underset{(44.28)}{0.039x} - \underset{(-14.92)}{0.22}, \qquad \bar{R}^2 = 0.993$$

(figures in parentheses represent t ratios).

28. Although there is no actual point of satiety in the case of sugar in urban areas, there is nevertheless a marked decline in the mpc as incomes rise.
29. Prais and Houthakker (1955).
30. On the maize-grinding and laundry soap markets in Kenya see respectively Stewart (1977) and Langdon (1975).

References

P. K. Bardhan (1974). 'On the Incidence of Rural Poverty in the Sixties', in P. K. Bardhan and T. N. Srinivasan (eds), *Poverty and Income Distribution in India*, Statistical Publishing Society.

C. Baron (1975). 'Sugar Processing Techniques in India', in Bhalla, A. S. (ed.), *Technology and Employment in Industry*, Geneva: ILO.

A. S. Bhalla (ed.) (1975). *Technology and Employment in Industry*, Geneva: ILO.

G. S. Chatterjee and Bhattacharya, N. (1974). 'On Disparities in Per Capita Household Consumption in India', in Bardhan, P.K. and T. N. Srinivasan (eds), *Poverty and Income Distribution in India*, Statistical Publishing Society.

W. R. Cline (1972). *Potential Effects of an Income Redistribution on Economic Growth*, New York: Praeger.

W. R. Cline (1975). 'Distribution and Development: A Survey of Literature', *Journal of Development Economics*, vol. 1.

V. M. Dandekar, and N. Rath (1971). 'Poverty in India – I: Dimensions and Trends', *Economic and Political Weekly*, January.

V. M. Dandekar and N. Rath (1974). 'Right to Gainful Work', in P. K. Bardhan and T. N. Srinivasan, (eds), *Poverty and Income Distribution in India*, Statistical Publishing Society.

A. Goldberger (1964). *Econometric Theory*, Chichester: Wiley.
Government of India (1959–60). *Sugar in India.*
Government of India (1965). *Report of the Sugar Enquiry Commission*, Delhi.
Indian National Sample Survey (1961–2). *Consumer Expenditure Survey*, 17th Round.
Indian Sugar Yearbook (1969–70), p. 119.
ILO (1970). *Towards Full Employment*, Geneva.
S. Langdon (1975). 'Multinational Corporations' Taste Transfer and Underdevelopment: A Case Study from Kenya', *Review of African Political Economy*, January–April.
P. C. Mahalanobis (1958). 'Science and National Planning', *Sankhya*, vol. 17.
J. E. Meade (1976). *The Just Economy*, London: Allen & Unwin.
D. Morawetz (1974). 'Employment Implications of Industrialization in Developing Countries: A Survey', *Economic Journal*, September.
F. Paukert, J. Skolka and J. Maton (1976) *Redistribution of Income, Patterns of Consumption and Employment*, Geneva: ILO.
S. J. Prais and H. S. Houthakker (1955). *The Analysis of Family Budgets*, Cambridge University Press.
K. Prasad (1963). *Technological Choice Under Development Planning*, Popular Prakashan.
S. C. Roy (1951). *Monograph on the Gur Industry of India*, Indian Central Sugarcane Committee, Kanpur.
A. K. Sen (1973). *On Economic Inequality*, Oxford: Clarendon Press.
F. Stewart (1977). *Technology and Underdevelopment*, London: Macmillan.

APPENDIX TO CHAPTER 3
A GEOMETRICAL REPRESENTATION OF THE RELATIONSHIP BETWEEN INCOME REDISTRIBUTION AND CONSUMPTION*

In this note we present a simple geometrical framework in which it is possible to examine the impact of an income redistribution in favour of the poor on the consumption of a commodity. (In Chapter 3 the commodities concerned were gur, sugar and a composite of the two sweeteners, but the analysis can be generalized.) It is also possible within this framework to investigate the question of which distribution of income produces the maximum consumption of a particular commodity (or, more generally, consumption expenditure as a whole).

In Figure 3A.1 over the range $Y_1 Y_2$ the curvilinear consumption function C is strictly concave since the function is such that linear

---

* From J. James (1976) *Technology, Products and Income Distribution: A Conceptualisation and Application to Sugar Processing in India*, ILO, World Employment Programme Research, Working Paper 61.

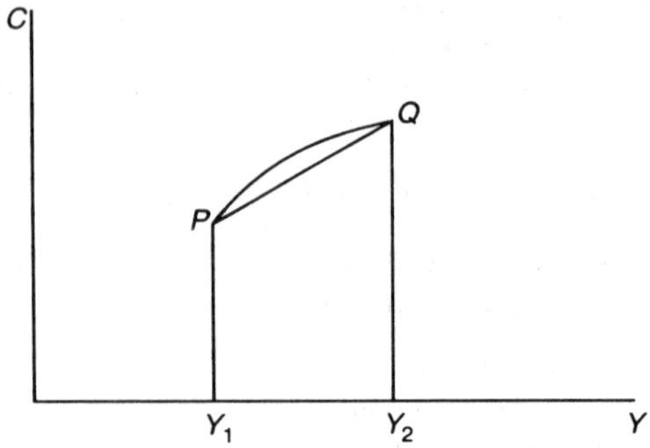

Figure 3A.1 Strictly concave consumption function

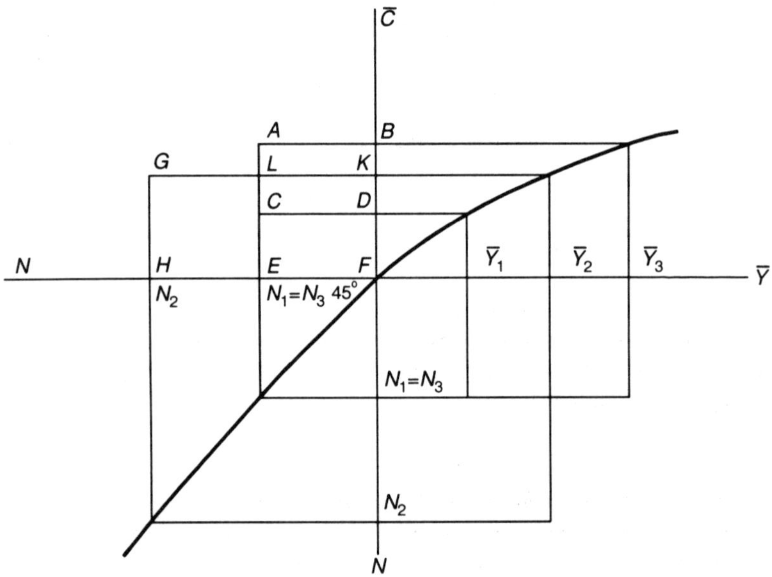

Figure 3A.2 Redistribution and consumption

interpolation between any two points always gives lower values for C than the function itself, i.e. between $Y_1\,Y_2$:

$$[SC(Y_1) + (1 - S)C(Y_2)] < C[S(Y_1) + (1 - S)(Y_2)]$$
$$(\text{where } S < 0 < 1)$$

any averaging of incomes between Y_1 and Y_2 must therefore raise total consumption.[1] In the case of strict convexity, averaging of incomes will reduce total consumption since

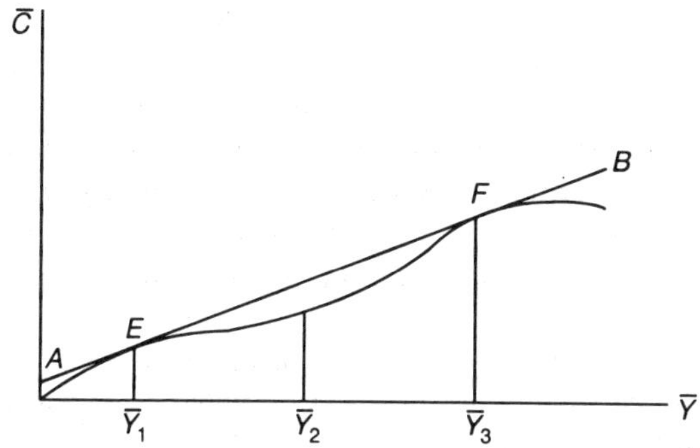

Figure 3A.3 Income distribution and maximum consumption

$$[SC(Y_1) + (1 - S)C(Y_2)] > C[S(Y_1) + (1 - S)(Y_2)]$$

and linear interpolation yields higher values than the function. Where the function is linear the left-hand side of the inequalities is equal to the right-hand side so that averaging produces no effect on consumption.[2]

Using these concepts the process of income redistribution and its effect on consumption can be illustrated by the four-quadrant diagram of Figure 3A.2.

In Figure 3A.2 the lower right-hand quadrant shows a normal distribution (assumed for simplicity) with N representing numbers of individuals at each income level $\bar{Y}$ (where $\bar{Y}$ represents the average total per capita income for each income class to make the analysis correspond to grouped consumption and income data), and $N_1 = \frac{1}{2}N_2$. The upper right-hand quadrant shows the estimated functional relationship between $\bar{Y}$ and $\bar{C}$ (where $\bar{C}$ is the average per capita consumption expenditure for each group). The lower left-hand quadrant merely uses the 45° line to transpose the values of N to the horizontal axis and the last quadrant shows total consumer expenditure, i.e. $\sum_{i=1}^{n} \bar{C}_i \bar{N}_i$.

We assume for the sake of simplicity a community of three income groups so that in the pre-redistribution situation total income (Y) equals $N_1\bar{Y}_1 + N_3\bar{Y}_3 + N_2\bar{Y}_2 (N_3 = N_1)$. Using the strictly concave consumption function it can be seen in the upper left-hand quadrant that total consumption equals the sum of the rectangular areas *CDEF* + *ABEF* + *GKHF* or $N_1\bar{C}_1 + N_3\bar{C}_3 + N_2\bar{C}_2$.

The society is now made totally egalitarian by transferring $(\bar{Y}_3 - \bar{Y}_2)$ to $\bar{Y}_1$ so that each group has the mean income $\bar{Y}_2$. Total income

$2 \times (N_2\bar{Y}_2)$ is equal to $N_1\bar{Y}_1 + N_3\bar{Y}_3 + N_2\bar{Y}_2$. Total consumption equals $2 \times (GKHF)$ and the increase is seen to equal *LKCD -ABLK* (since *GLHE* equals *LKEF*). It can also be seen that given the strictly concave consumption function, this configuration of incomes yields the maximum consumption expenditure since, for any movement from the mean income, *ABLK* < *LKCD*. Analogous but converse results follow for a strictly convex function. With a linear (convex) function no change occurs. With functions not uniformly either strictly convex or concave, no a priori conclusions can be drawn. The configuration which maximizes total consumption is, however, derivable.

Along the line segment *AB* of Figure 3A.3 it is true by convexity that

$$\bar{Y}_2 N = S(\bar{Y}_1)N + (1 - S)\bar{Y}_3 N \quad (0 < S < 1)$$

where $\bar{Y}_2$ is average income and N is the number of individuals with such average income. Maximizing consumption in this case involves notionally moving those with average incomes to $\bar{Y}_1$ and $\bar{Y}_3$ in the proportions S and (1-*S*) (which can be solved from the last equation). This result follows from the fact that below *AB* consumption can be improved by, as it were, shifting those at the mean to the local maxima *E* and *F*. Should redistribution have as its aim the maximum consumption of a commodity or commodities some would have to be made poorer and some richer if the curve were shaped as in Figure 3A.3 with strict convexity and concavity over part of the income range.

Table 3A.1 Consumption functions

Name	*Form*	*Slope*	*Elasticity*
1. Linear	$y = a + bx$	b	bx/y
2. Semi-logarithmic	$y = a + b \log x$	b/x	b/y
3. Hyperbolic	$y = a - \frac{b}{x}$	b/x^2	b/xy
4. Double-logarithmic	$\log y = a + b \log x$	by/x	b
5. Log-reciprocal	$\log y = a - \frac{b}{x}$	by/x^2	b/x
6. Quadratic	$y = ax^2 + bx + c$	$2ax + b$	$\frac{2ax+b}{y/x}$

where:
y = expenditure on the specific commodity
x = total income
$\frac{dy}{dx}$slope
$\frac{(xdy}{ydx)}$ = elasticity.

We may now examine the properties of the widely used Engel curves in order to discover their implications for consumption when used in simulation exercises. Table 3A.1 sets out these curves in terms of derivatives and elasticities.

We may classify these by the behaviour of the slope (mpc).

1. *MPC falls monotonically:*
   (a) semi-logarithmic function
   (b) hyperbola
   (c) quadratic function with coefficient of x^2 negative
   (d) double-logarithmic with $b > 0 < 1$.
2. *MPC rises monotonically*:
   (a) quadratic function with coefficient of x^2 positive
   (b) double-logarithmic with $b > 1$.
3. *MPC constant*:
   (a) linear function.
4. *MPC rises (falls) and then falls (rises)*:
   (a) log-reciprocal.

Notes

1. In terms of Lorenz curve analysis it may be noted that only a lower Lorenz ratio with non-intersecting curves constitutes an unambiguous move towards the average, i.e. a redistribution which moves everyone towards the average or some towards the average and none away from it.
2. An analogous use is made of strict concavity by Sen in the context of utilitarian welfare economics showing how, with a strictly concave welfare function, averaging raises social welfare. (A.K. Sen (1973). *On Economic Inequality*, Oxford: Clarendon Press.)

4 From Products to Functionings: New Medicinal Drugs in Developing Countries

(with P. Parthasarathi)

Set partly in the context of the debate about the economic impact of multinationals on developing countries, numerous attempts have been made to study the welfare effects of new products on the inhabitants of those countries. For the most part, these studies are based on a form of analysis (developed primarily by Lancaster) that focuses on the specific characteristics embodied in new products.[1] Those combinations of characteristics are then evaluated from the point of view of their appropriateness to household groups differentiated, most frequently, by levels of income. Because they are mainly produced in developed countries, new products, so the argument usually runs, contain a high proportion of 'high-income' characteristics that make them unconducive to the satisfaction of basic needs in developing countries. Or, put another way, the conclusion is typically that new products tend to confer their benefits disproportionately on rich rather than poor consumers in developing countries.

From the point of view of individual welfare, however, the problem with this product characteristics oriented approach is that, as Sen has pointed out, it constitutes only part of the link between consumption and individual well-being and thus between new products and well-being.[2] What also matters, according to Sen, is 'what the person succeeds in *doing* with the commodities and characteristics at his or her command. For example, we must take note that a disabled person may not be able to do many things an able-bodied individual can, with the same bundle of commodities'.[3] This comparatively neglected aspect of the relationship between consumption and welfare is what Sen calls 'functionings' of persons, and our purpose here is to show how his approach can be analysed in relation to one particular type of new product, namely, medicinal drugs. There are two reasons why we have selected this particular product. One is that the link between the characteristics embodied in modern medical drugs and their ultimate impact on health functionings is to a large extent objectively measurable. Indeed, there is a large medical literature which is available for this purpose, but which, until now, has gone largely unused in economic development or welfare economics, on account of what Sen refers to as a lamentable 'reluctance to go into medical matters'.[4] The second reason is that the functionings to which modern drugs give rise in developing countries exhibit more individual variation than most, if not all, other new products and this variation needs to be understood if one is to lessen the gap between those whose functionings are enhanced and those whose functionings are left unchanged or affected adversely.

Following a brief description of how the concept of functionings fits into Sen's overall framework for analysing the welfare effects of consumption, we show in the section thereafter why there is so much scope for individual variation in the impact of modern drugs on health functionings. Next, we discuss what factors determine whether that impact is positive, negative or negligible in particular cases. The final section examines the relationships between the variables thus identified.

FUNCTIONINGS IN SEN'S MODEL

As set out in his *Commodities and Capabilities*,[5] Sen's analysis is concerned with the relationships between commodities, characteristics (of those commodities), functionings and capabilities. It is not our intention here to describe all these complicated relationships in detail. Rather, what we wish to convey in simple graphical terms is the meaning of Sen's statement that,

> A functioning is . . . different both from 1) having goods (and the corresponding characteristics), to which it is posterior, and 2) having utility (in the form of happiness, resulting from that functioning), to which it is, in an important way, prior.[6]

The 'posterior' relationship, involving the connection between goods and characteristics – is familiar from Lancaster's framework and is shown in the top left-hand panel of Figure 4.1. Indeed, as noted above, it is mainly this type of diagram that has been used in one form or another to analyse the welfare effects of new products. As was also noted previously, however, in Sen's approach a second element needs to be added to the analysis in characteristics space, given his view that the welfare effect of a given bundle of characteristics depends on what is actually done with them. This second basic element of Sen's model is reflected in the top right-hand panel of Figure 4.1, which, for the sake of simplicity, deals with the relationship between individual functionings and just one of the two characteristics used in the first panel.

Consider for example the two goods X and Y, which, we assume, contain the amounts OP and OR of Characteristic 1 respectively. The extent to which these two amounts actually contribute to individual functionings, however, depends on the slopes of the lines connecting P and R to the horizontal axis in the top right-hand panel. Sen refers to

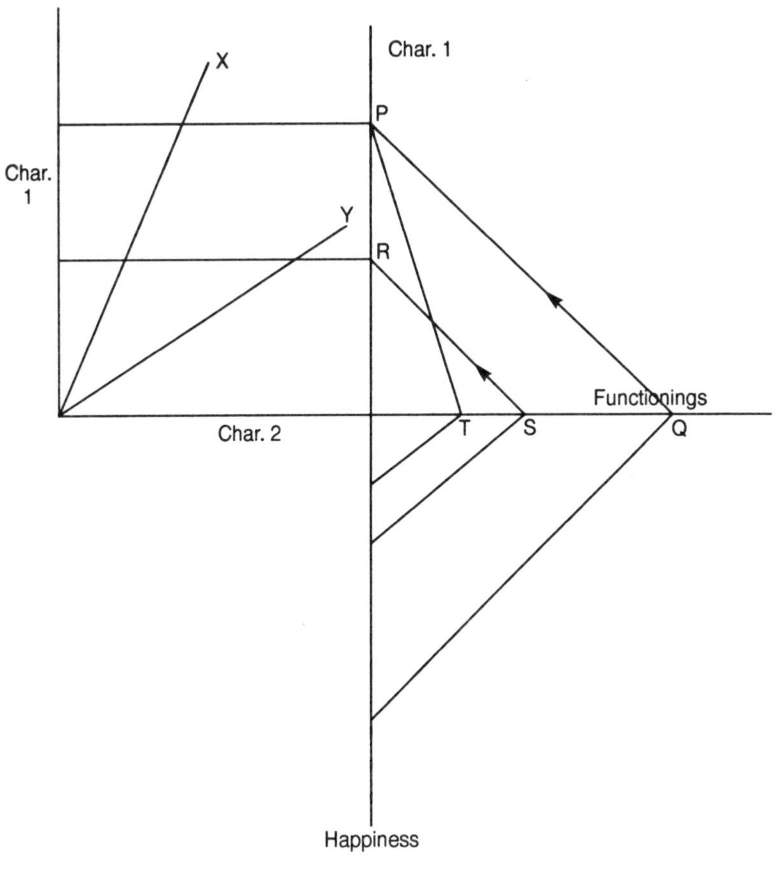

Figure 4.1 Functionings in Sen's approach

these types of functions as 'personal utilization functions' and they depend on a variety of both 'choice' and 'non-choice' factors. The impact of a given amount of nutritional characteristics on nutritional functionings, for example, depends on a person's metabolic rate (a non-choice factor) and the precise way in which he or she prepares the food (a choice factor).[7] Thus, depending on the particular combination of choice and non-choice factors in our example, the utilization functions associated with good X may be more or less favourable to that individual's functionings (PQ for example as opposed to PT). In the latter case (and assuming that the utilization function associated with good Y happens to have the same slope as PQ), then this good will have a more favourable impact on the person's functionings than good X, even though the latter embodies more of the characteristic in question. By means of a similar line of argument, one can readily show that varying utilization functions produce correspondingly different effects on the functionings of two different consumers of the same amount of product characteristics.

The last element of Sen's model that can be captured in our simple graphical treatment is shown in the bottom right-hand panel of Figure 4.1. It reflects the fact that functionings are not an end in themselves but should be regarded instead as 'prior' events in the achievement of happiness. Sen refers to the function connecting these two variables as the 'happiness function' of an individual and, in the illustrative example of Figure 4.1, this function is assumed to be such that happiness stands in a simple linear relationship to the achieved degree of functionings.

In the rest of this chapter, we shall primarily be concerned with the nature of the utilization functionings connecting drug products and health functionings (that is, with the upper-right panel in Figure 4.1). First, however, we need to examine why the scope for variation in individual functionings is so considerable in the case of these particular products.

MODERN DRUGS AND THE SCOPE FOR INDIVIDUAL VARIATIONS IN FUNCTIONINGS

Prior to the introduction of synthetic chemicals in the 1920s, the formulation of drugs relied mainly on natural plants and though the latter are still important in certain fields, the former are currently the major source of new drugs. Among the changes wrought by this transition was not only a dramatic increase in the number of drugs but also a striking change in the characteristics that they embodied. Most important from our point of view is the change that occurred in the relationship between the therapeutic efficacy of drugs and their potential toxicity. 'Unlike many of the older agents', that is to say, 'which were by and large relatively ineffective but also relatively safe, the new drugs were often far more powerful – and far more toxic. . . . Physicians were accordingly faced not only with an astounding array and complexity of drugs, but with drugs possessing an equally astounding potential for curing – or for killing.'[8]

Expressed slightly differently, what occurred was a greatly enlarged potential for variation in the impact of drugs on individual health functionings. Consider, for example, Figure 4.2, which contains characteristics on the vertical axis and functionings on the horizontal axis and which compares the case of a drug produced in (say) 1900 with a drug designed to treat the same illness in (say) 1990. Because the latter possesses potentially more of both efficacy and toxicity than the

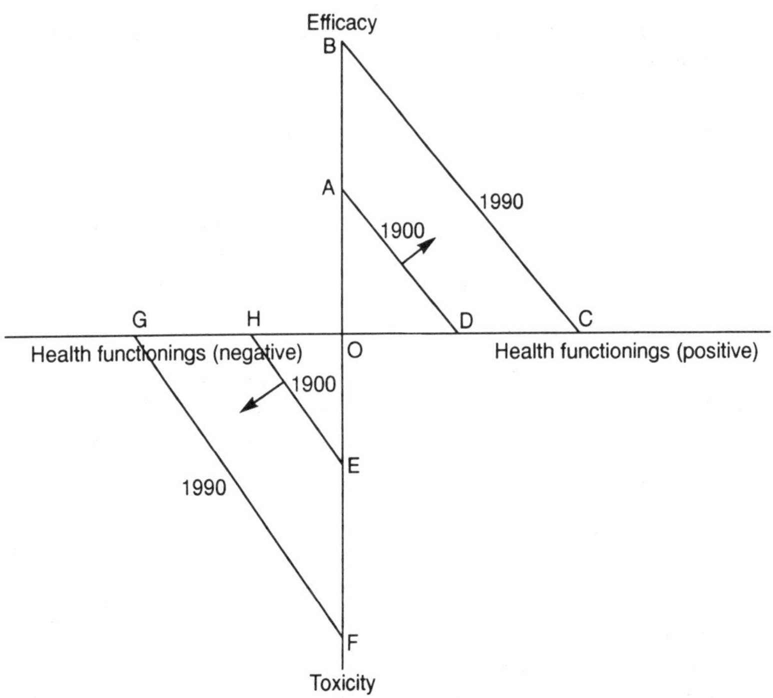

Figure 4.2 Drug efficacy, toxicity and health functionings over time

former (OB and OF versus OA and OE respectively), it increases the range of health functionings from HD (in the case of the 1900 product) to GC. The remainder of the chapter is concerned with the factors that determine whether the consumer of a modern drug product in a developing country ends up at point G, point C or some intermediate position. Our discussion is based on a distinction drawn by G.A. Cohen between what a good 'does' for a person as opposed to what the person does with the good.[9] This distinction is important for policy purposes because to a considerable extent it overlaps with the distinction between factors that are and those that are not amenable to policy interventions designed to alter the relationship between consumption and individual functionings.

'What Drugs Do For People'

One reason why the individual has little or no control over the impact of drug consumption on his or her functionings is that this impact is heavily conditioned by factors innate to the individual. Table 4.1 lists the most important of those factors and provides examples of the mechanisms through which they operate.

Many of the mechanisms and examples given in Table 4.1 are, of course, common to individuals in developed as well as developing countries. Other cases, however, are more prevalent in, or entirely specific to, the latter group of countries. Some of the variations listed in the table can be predicted (and, where unfavourable, avoided), whereas other variations are more idiosyncratic and less easy to predict. Predictability, one should emphasize, also depends on the nature of the drug itself. In the case of Lomotil, for example (a drug that is sometimes given in the treatment of infant diarrhoea), the 'narrow range between therapeutic and toxic doses, and also the possibility of a child being abnormally sensitive... may account for the severe toxicity sometimes seen with low dosage.'[10] The same is true of drugs such as digoxin (which are used in the treatment of heart disease), in that 'The therapeutic level is 60 per cent of the toxic level. The toxic level, moreover, may vary considerably between one patient and another.'[11]

The main conclusion to be drawn from Table 4.1, however, is that even without any differences in the way they behave, individuals can derive dramatic differences in health functionings from consumption of exactly the same therapeutic characteristics. Some individuals, that is to say, may end up near the one extreme point on the horizontal axis of Figure 4.2, while other individuals, behaving in the very same way, end up closer to the other extreme.

Table 4.1 Source of individual-specific variations in health functionings following drug usage

Source	*Mechanisms/examples*
Age	'The main reason that age affects drug action is that drug metabolism and renal function are less efficient in babies and old people, so that with some exceptions, drugs tend to produce greater and more prolonged effects at the extremes of life.' (Rang, Dale and Ritter, 1995, p. 785)
Genetic factors	'Genetic variation is an important source of pharmacokinetic variability' (i.e. variability due to altered handling of drugs by the body, which leads to differing concentrations of the drug at the site of action). (Rang, Dale and Ritter, 1995, p. 789) Genetic variation can also cause differing physiological responses to the *same* drug concentration and some such responses may often be harmful (e.g. allergic individual responses even at low dosages). Glucose-6-phosphate dehydrogenase (G-6-PD) is a genetic disorder that is widespread in Africa, the Mediterranean, the Middle East and Southeast Asia. Certain types of drugs can produce hemolysis (a disorder of the blood) in G-6-PD-deficient individuals (some anti-malarial drugs, for example, produce this condition among Africans, Chinese and Thais (Weatherall, 1990)), though it is generally well tolerated in the general population.
Disease	'Disease can cause altered handling of drugs by the body (pharmacokinetic variation) and/or altered sensitivity to drugs' (Rang, Dale and Ritter, 1995, p. 42). Vaccines given to those suffering from protein-energy malnutrition may be ineffective, or less effective than vaccines given to those with adequate amounts of these nutrients.[12]
Physiological individuality	Pregnancy, weight (the therapeutic function of a drug may often be inversely related to the weight of patients), sex (female patients tend to react more strongly to most medications and to suffer from a higher incidence of adverse effects).[13]
Psychological individuality	The placebo effect: 'It has been found that a placebo can potentiate, attenuate, or negate the active ingredients in a drug.' (Wickramasekera, 1985, p. 256)

Environmental determinants of inter-individual variations in health functionings

The examples that have just been cited are not the only ones to be considered under the heading of 'what drugs do for people'. For there are also environmental factors that, on the one hand, influence the relationship between drug consumption and health functionings and on the other hand are unamenable to individual influence. 'The effects of medications may be markedly altered', for example, 'if the physiological equilibrium is unbalanced by altitude, cold, heat, humidity, sunlight, vibrations, and other stresses'.[14] But what comes perhaps most readily to mind in the current developing country context are the large differences between and within countries in the extent to which there is now resistance among the population to anti-malarial and antibiotic medications.[15] Since these medications are among the most commonly prescribed in the Third World, such environmental differences as these clearly form an important part of any study that is concerned with the question of why given drugs are apt to elicit such marked differences in health functionings among individual patients in the Third World.

What is less clear, though, is whether the drug resistance problem forms part of the category 'what drugs do for people' or whether it should be placed under the heading 'what people do with drugs'. On the one hand, there are a number of studies which suggest that resistance is closely related to the way in which the drugs themselves are used (and to this extent the issue should be dealt with below when we consider how functionings vary with drug usage). 'Drug pressure', concludes one such study of malaria resistance in Africa, 'is the main determinant of the rate of spread and development of resistance.'[16] Similarly, there is evidence of a direct and close relation between antibiotic use and antibiotic resistance in certain other parts of the Third World.[17]

On the other hand, it also seems unlikely that drug usage patterns alone can account for the entire resistance problem. 'To some extent', for example, the global incidence of genes that are resistant to antibiotics 'may relate to the initial presence of the resistance determinant in a particular area'.[18] Recent malaria research, moreover, indicates that 'natural geographic variation in parasite mating habits may also play a role'.[19] That variation, it seems, has to do with differential degrees of inbreeding among malaria parasites in different parts of the Third World, with more pronounced inbreeding patterns tending to produce more acute resistance. Those patterns, in turn, appear to be heavily influenced by differences in local rates of transmission of the disease. In Tanzania, for example, the rate of transmission from mosquitoes to humans is relatively high: on average Tanzanians receive between 300 and 3,000 infected bites per year. 'A single human host often carries many different strains of malaria at once, and mosquitoes are likely to take up mixed strains, allowing frequent cross-mating.'[20] In Papua New Guinea, on the other hand, 'the transmission rate of malaria is 10 times lower, which means more mating between parasites of the same strain'.[21]

'What People Do With Drugs'

Under this heading one needs to consider (1) whether a given medication is capable (in principle) of effectively treating the illness or disease from which the patient is suffering (viral infections for example do not respond to antibacterial drugs) and (2) whether the medication is used in a manner that makes the maximum therapeutic contribution to the treatment of that particular illness or disease. This second issue, of course, only becomes relevant if the first condition is satisfied, the likelihood of which itself depends on the accuracy of the diagnosis, as well as the suitability of the medication that is chosen to treat the particular problem. (Here again the issues are sequentially related: a suitable drug can only be chosen once an accurate diagnosis has been made. Note too that the term 'suitable' is being used here in a clinical sense and not in the sense of whether the characteristics embodied in a drug product are appropriate to persons with different incomes.)

Given the choice of a product that *is* capable of treating a particular illness effectively, a range of factors bearing on the use of the product then determines whether this potential is actually realized, or whether other, possibly negative, impacts on health functionings will occur instead. Of these types of variable, the following three, somewhat overlapping categories, are perhaps the most important.

1. *Compliance*: To a greater or lesser extent, the efficacy of all drugs is dependent upon patient compliance with factors such as dosage (how and how frequently the drug is to be taken, at what times of the day, for what length of time, etc.); storage (under certain conditions of temperature, light, etc. drugs tend to lose their efficacy, a problem that assumes particularly serious proportions in tropical conditions[22]); and simultaneous use with other drugs (see below).

For some drugs, however, compliance is more of a problem than others. It is an especially problematic issue in the treatment of tuberculosis, for example, because of the length of the treatment (6 – 12 months) and the need to take several drugs.

2. *Interactions*: 'Drug products may be made dangerously toxic or may be partially or wholly prevented from eliciting the desired response in patients because of interactions with other substances.'[23] Many drugs, for example, 'interact adversely with certain other drugs, food constituents, and other chemicals'.[24] Such interactions, one should emphasize, are not confined only to prescription drugs; they frequently occur with 'over-the-counter' medications as well.[25]
3. *Complementary inputs*: We are referring here not only to the physical inputs that are often required for the effective use of modern drugs (such as refrigeration, measuring instruments, etc.) but also to the human capital inputs that are required for this purpose, e.g. literacy, consumption skills.

Even in the developed countries, for whose inhabitants the vast majority of modern drugs are specifically designed, there is variation in what people do with drugs and so too therefore is there variation in the impact of those drugs on individual health functionings[26] (see below, for example, on differing rates of compliance in developed countries). In the Third World, even more marked variations may be expected in the relationship between drug consumption and individual health functionings. For while a relatively small group of consumers in those countries tends to derive something akin to the maximum therapeutic benefits from modern drug products, in the vast majority of cases there are a number of factors that make such an outcome extremely unlikely. We can best consider these factors in relation to the variables we have already described in this section, namely, diagnostic accuracy and drug suitability, patient compliance, drug interactions and complementarities in drug use.

Diagnostic accuracy and drug suitability

To a far greater extent than in developed countries, modern drug use in the developing world is unaccompanied by medical supervision. In large part, this is because drugs which in developed countries are available only on prescription, can often be freely purchased without any such restriction. But it also reflects the fact that for many poor consumers in the Third World, low incomes give rise to a conflict between purchasing medications on the one hand and medical services on the other.

As is well illustrated by a study of drug use among rich and poor households in Brazil, this conflict is usually resolved in favour of the medication rather than the medical services. In particular, 'the number of medicines in use in the peripheral shanty towns and in the best suburbs... did not vary much'.[27] However, 'The most important difference between the rich and poor was that the majority of the poor (52 per cent) obtained their medicine without a prescription, which the rich usually had. Further analysis confirms that the real difference between rich and poor in Brazil is in their access to doctors.'[28]

A similar picture emerges from a study of drug use patterns among low-income rural households in the Philippines, who, it seems, 'rely more on western medicines than on western doctors'[29] and who use such medicines 'even in minor self-controlling ailments'.[30] (All told, in this sample, modern pharmaceuticals, prescribed and non-prescribed, were used in about 50 per cent of the cases where no professional was consulted.) The problem with this pattern of usage is that in many cases, the drugs thus selected for home use turn out to be a highly inappropriate form of treatment, as when, for example, anti-diarrhoeal combination products are used against infant diarrhoea, even though they are generally not recommended for this purpose, not only because they expose patients to unecessary risks but also because they contribute to the drug resistance problem noted earlier.[31]

This comparative lack of correspondence between drug use and medical supervision is not, however, the only factor that makes the prospect of effective treatment comparatively unlikely in developing, as opposed to developed countries. For even when medical supervision *is* available, 'Much – and perhaps everything – depends on whether the product is described in the labeling and promotion so that the physician can order the right drug for the right patient, in the right amounts and at the right time, and so that he can given each patient the proper instructions for use.'[32] In these respects, too, unfortunately, there is good reason to suppose that patients in the Third World tend to be severely disadvantaged in comparison with their counterparts in the rich, industrialized countries. Much of this difference, it seems, has to do with what 'drug companies tell physicians in industrialized nations',[33] as opposed to 'what they say in the developing countries of the Third World'.[34] Consider, in particular, what one very detailed study of multinationals operating in a wide range of developing countries, has to say on the topic. Thus,

> *Many products ousted from the market in such industrialized nations as the United States and Great Britain, or never approved for marketing, are readily available and widely promoted in the Third World.... With many important products, the dangers of serious or lethal side effects are minimized, glossed over, or totally ignored.... With many of these drugs, claims of effectiveness are wildly exaggerated.*[35]

Possessed thus of information that is systematically inferior to what is propagated by the same firms in the developed countries, physicians in the Third World are correspondingly less likely to provide their patients with appropriate diagnostic and other care.

Compliance and drug interactions

Rates of compliance with recommended dosages appear to be extremely low in developing countries. One Indian study, for example, suggests that out of the total number of antibiotics purchased without a prescription, only 11 per cent were taken in the correct dosage for a long enough period of time (and as many as 42 per cent of the drugs were taken for less than a day).[36] And even when the antibiotics were purchased from a medical practitioner, they were adequately prescribed in only a quarter of the cases. A separate study of Indian patients found a similar pattern: 'among the self medicated, 30% purchased antibiotics for less than 1 day and only 18% purchased them for 5 days...on the doctor's prescription only 40%...were for a full course (minimum five days).'[37]

To a degree, findings such as these reflect some of the same factors that cause differences in patient compliance in developed countries, such as ambiguities in instructions or motivational differences among patients which affect their willingness to comply.[38] After all, even in those countries where conditions are comparatively favourable, rates of compliance with usage requirements in the case of antibiotics, are said to be no more than 50 per cent.[39] To a probably greater degree, however, the findings reported in the previous paragraph reflect problems that are peculiar to, or more pronounced in, the developing countries themselves. In those countries, for example, the poorest groups of patients often purchase one or two loose tablets from an open bottle whose package insert has already been removed; physicians often fail to provide any information about dosage requirements and such information, even when it is available, is often in a form that is inaccessible to poorly educated consumers.[40]

Because of the lack of information with which they are associated, these same factors, moreover, tend to raise the probability of unfavourable drug reactions in developing, as compared with developed, countries. Against this, one could argue that patients in developing countries use fewer drugs simultaneously, thus lessening the likelihood of unfavourable drug interactions. Yet, while this argument may hold in some cases, it is well to emphasize that 'in many developing countries up to six or seven drugs may be prescribed for conditions in which one or two would do'.[41]

Complementarities in drug use

If it is to be effective, modern drug use often depends on the availability of certain inputs, some of which (such as refrigerators[42] and sterilized syringes) take the form of physical capital, while others are better described in terms of the human capital that is necessary to read and understand dosage requirements, contra-indications, etc.

In developed countries and those parts of the Third World with relatively high levels of income and education, these requirements are widely met and the translation of drug consumption into health functionings is consequently (from this point of view) not impaired to any significant extent. Among the poorest households in developing countries, by contrast, the all too frequent absence of the inputs we have just described means that a given drug will tend to have a less beneficial and sometimes even a highly unfavourable impact on health functionings (in much the same way that an absence of refrigerators and clean water has so adversely influenced the effect of infant formula (baby milk) in developing countries).

WHAT PEOPLE DO WITH DRUGS AND HEALTH FUNCTIONINGS

We suggested earlier that the role of functionings in Sen's approach could be explained in simple graphical terms. We now show that this same framework can be used to illustrate and compare the mechanisms through which differences in drug usage behaviour influence individual health functionings.

As before, the issue is approached in terms of how the relationships in characteristics space (the left-hand panel of Figure 4.3) are mapped into differences in functionings (the right-hand panel). Consider to begin

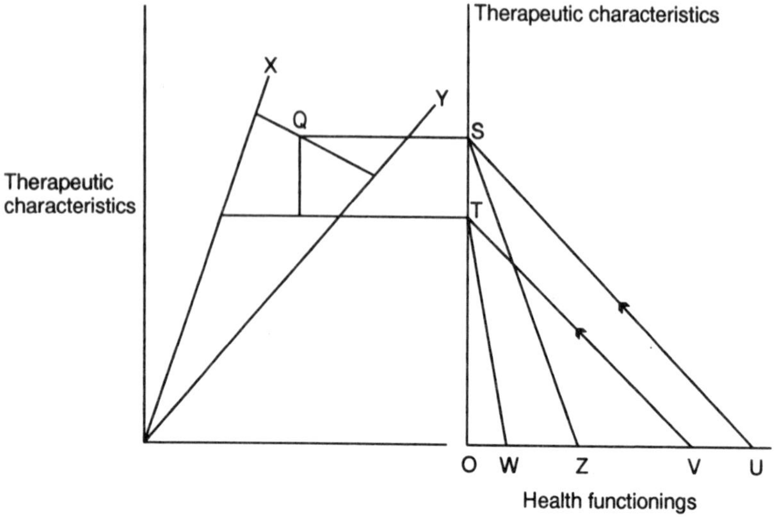

Figure 4.3 Drug usage, characteristics and functionings

with the 'bench-mark case' where the individual derives the maximum therapeutic benefits from the drug products that he or she consumes (this begins in Figure 4.3 at point Q, which is translated, we assume, into functionings by the line SU). Deviations from this outcome occur either because the individual derives less than OS of the therapeutic characteristics embodied in the product, or because that amount of characteristics translates itself into functionings via a less favourable function than SU (Of course, a combination of these categories is also possible). Into the first category would fall cases, where, for example, the individual fails to comply with drug usage requirements (by taking the drug at say the wrong time of day), or where negative interactions between two drugs occur. In Figure 4.3, we assume that events such as these cause the individual to derive only the amount OT of therapeutic characteristics and hence to derive proportionately less in terms of health functionings (assuming, as we are, that TV runs parallel to SU).

The second category includes the case where the individual takes the drug entirely as recommended, but because it is clinically unsuited for the particular illness from which she is suffering, no positive impact on health functionings emerges (and the function connecting therapeutic characteristics and health functionings then coincides with the vertical axis). The problem of drug resistance can also be placed in this category to the extent that it is caused by factors other than inappropriate use patterns, for what then occurs is an inward movement over time of the line SU (to say SZ), *with no reduction at all in the amount of available therapeutic characteristics.* To some degree, though, resistance is itself partly the result of misuse, and the deviation from the 'ideal' outcome should then be depicted as a combination of fewer characteristics *and* an inward shift of the line SU, resulting, ultimately, in a point such as W on the functionings axis.

AN INTEGRATED POLICY FRAMEWORK

At various points we have suggested that the impact of drug consumption on individual health functionings depends on variables that are sequentially related. From a policy point of view, it is useful to present these types of relationships in the graphical form of a 'tree', each of whose forks represents the outcome of a particular stage in the process by which drug consumption is converted into health functionings.

At each of the first three forks in Figure 4.4 (collectively representing the category 'what people do with drugs'), there is an outcome which is necessary (but not sufficient) for drug consumption to engender an

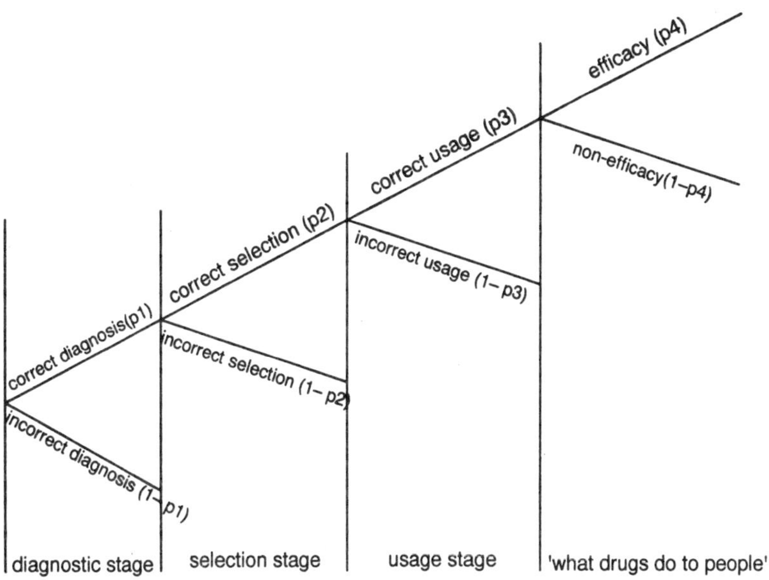

Figure 4.4 Sequentiality in the determinants of health functionings

improvement in the health functioning of an individual and an outcome which prevents this from occurring (regardless of the stage when it occurs and regardless of whether the upper fork has a very high probability of occurring in any subsequent stages). Framing the issue this way is of course something of an oversimplification since in reality there can be degrees of satisfaction with variables such as correctness of usage; the point of the 'tree' diagram however is less to capture the details of the process and more to emphasize its sequentiality. The last fork, which is labelled efficacy, corresponds to what we referred to earlier as 'what drugs do to people' and it shows that even when the first three outcomes are favourable, there may still be some probability that a drug will fail to have the desired effect on health functionings.

If we assume that all of the variables in the figure are independent of one another, then the likelihood that a given medication will elicit the desired response (what in relation to Figure 4.3 we referred to as the 'bench-mark case'), becomes a multiplicative function of the four individual probabilities, p1, p2, p3 and p4.[43] We have argued that the first three probabilities tend to be low in developing rather than developed countries and among low- rather than high-income consumers in those societies. Not all medications, however, conform to this typical pattern and the exceptional cases can also be explained in terms of the probabilistic framework we have just described. Measles vaccines, for example, are said to have an efficacy of 85 per cent in that they reduce the incidence of measles by that percentage and since the remaining probabilities are equal to 100 per cent (by definition), the overall probability is also 85 per cent.[44] ORT (oral rehydration therapy) also has a very high efficacy (as measured by the decrease in case fatality rates) and unlike the vaccine it is susceptible in principle to probabilities of less than 100 per cent in the other variables. Field data suggest, however, that diagnostic accuracy is also very high (approximately 95 per cent) and patient compliance is said to be in the range of 60 to 80 per cent.[45] Thus, with efficacy estimated to be 95 per cent, as already noted, the combined probabilities are in the region of 54–72 per cent.

It is worth emphasizing that outcomes such as these are only possible when all upper branches of the tree exhibit relatively high probabilities (that is, when there are no significantly weak links in the chain of sequentiality). This point is worth emphasizing because of its implication that the major policy challenge for the vast majority of other drug products is to simultaneously raise the existing low probabilities at all stages of the process by which drug consumption is converted into health functionings in developing countries.

CONCLUSIONS

Most existing discussions of the welfare effects of new products in developing countries are based on the appropriateness to different household groups of the characteristics that are embodied in those products. Basing our argument on Sen's notion of functionings, we take the view that this approach constitutes only part of the true relationship between consumption of new products and individual well-being.[46] What also matters is what the individual is capable of doing with the characteristics at his or her disposal. This additional consideration, we believe, is especially relevant to the case of modern drug products in developing countries, since there is so considerable a degree of variation in the way that different individuals respond to consumption of the very same medication. Our main concern in this chapter has been to examine the origin of these pronounced variations in individual health functionings and thereby better to understand the actual welfare effects of drug products in developing countries. What we find is that some such variations are due to factors over which the individual consumer of a particular drug product has little or no control (and these factors were accordingly described as belonging to a category entitled 'what drugs do to people'). To a probably greater extent, however, the observed differences in health functionings appear to derive from what individuals actually do with the drugs that they purchase and it is in this respect (rather than in the factors belonging to the first category) that systematic differences between groups of households are seen to emerge. In particular, though our evidence is based on case studies from different countries we found that modern drugs tend to be least effectively used by low-income and poorly educated consumers in developing countries, who, indeed, often seem to derive the toxic rather than the therapeutic characteristics of those products. Because the variables that give rise to this conclusion are sequentially related, policy interventions need to be directed to all stages in the process by which drug consumption is converted into health functionings among these particular groups of consumers.

Because the subject of this chapter has been new medicinal drugs designed in the rich countries, it is worth relating our conclusions to the notion of inappropriateness, which, as noted above, is sometimes said to reside in the transfer of products from developed to developing countries. Our view of the relationship between these two issues is that misuse of medical drugs in developing countries can be conceived in part as an additional dimension of product inappropriateness; additional, that is, to the other and better-known dimensions of inappropriateness such as product characteristics that (because they are designed for average tastes and incomes in the developed countries) are unsuited to the much lower incomes of the vast majority of Third World consumers. For, as we see it, medicinal and other new products are designed to fit in not only with average tastes and incomes in the developed countries, but also with the available consumption skills, infrastructure and climatic conditions in those countries. New products, like technological innovations in general, are designed, that is to say, to form part of particular *technological systems* and where elements of those systems (such as medical supervision) are missing, product misuse is the probable outcome. To this extent there is a need for medicinal products to be redesigned for specific developing country conditions rather than the consumption systems prevailing in the rich countries. Drugs could be redesigned for example with shorter treatment regimes to reduce the need for scarce medical supervisory support[47] or with a lesser demand for expensive complementary inputs such as refrigerators. Or again, there is a need for medicines that deteriorate less rapidly under tropical climatic conditions.

What also needs to be emphasized, however, is that drug misuse in developing countries cannot *entirely* be ascribed to product inappropriateness. The reason is that even in the developed countries these products are improperly used – albeit to a lesser extent than in the Third World – and it appears that there are reasons for misuse common to both types of countries (such as, for example, the lack of sufficient patient motivation for full treatment compliance). Where such common factors prevail, it is these rather than issues of inappropriateness *per se* that need to be addressed by policy.

Our final observation has to do with implications for further research in the area of drug consumption and individual health functionings. In particular there is a need for research to show how the various different patterns of drug use and misuse we have described translate ultimately into aggregate differences in infant mortality, health, longevity and so on. (To what extent, for example, does the greater misuse of antibiotic medications in developing countries influence patterns of health and longevity in those as opposed to developed countries?) At present there is no such specific evidence known to us.

There are however a few village-level studies that illustrate the kind of evidence that is needed to make the quantitative connections between product use and aggregate human functionings. Results from a community-based trial in Bangladesh, for example, suggest that use of oral rehydration salts (ORS) can lead to a 79 per cent reduction in diarrhoea case fatality rates.[48] Other evidence comes from the use of infant formula in developing countries which poses issues and problems that are closely related to those associated with medicinal drugs (especially the lack of certain necessary elements of the consumption system as a whole). There are in particular several studies that seek to quantify the impact of formula use on rates of hospitalization due to severe diarrhoea and on infant mortality at the village level.[49] In each case, formula use is shown to have had a strongly negative impact on these indicators in contrast to breast-fed infants in the same villages who did far better. Since the use of infant formula in developed countries generally has few such negative effects, the sharply higher rates of hospitalization and mortality that are reported for the developing countries can to a seemingly large extent be ascribed to the different patterns of use which are known to exist among resource-poor mothers in those countries.

Notes

1. See for example James and Stewart (1981), Helleiner (1975), Jenkins (1988), and Stewart (1977).
2. Sen (1985).
3. Ibid., p. 10, emphasis in original.
4. Ibid., p. 46.
5. Ibid.
6. Ibid., pp. 10–11.
7. This distinction is made in Sen, ibid.
8. Silverman and Lee (1974) p. 6.
9. Cohen (1993).
10. Snyder et al. (1973) p. 47.
11. Silverman and Lee (1974) p. 275.
12. Silverman, Lee and Lydecker (1982).
13. Martin (1978).
14. Ibid., p. 103.
15. See for example Schapira (1990).
16. Ibid., p. 43.
17. As reported in *The Lancet*, 10 July 1982, p. 86: 'A direct relation was found between antibiotic use and the frequency and kinds of antibiotic faecal strains in human beings in Toluca, Mexico.'

18. Ibid.
19. Schmidt (1995) p. 1670.
20. Ibid.
21. Ibid.
22. One study undertaken in Sudan, for example, showed that a number of pharmaceutical preparations underwent serious deterioration and degradation under tropical conditions. See *Tropical Doctor*, July 1980.
23. Martin (1978) p. 10.
24. Ibid.
25. Silverman and Lee (1974).
26. Ibid.
27. Muller (1982) p. 110.
28. Ibid.
29. Hardon (1987) p. 287.
30. Ibid.
31. 'The result is treatment failure, high costs and deaths.' Ibid.
32. Silverman, Lee and Lydecker (1982) p. 13.
33. Ibid, p. 14.
34. Ibid.
35. Ibid, p. 85, emphasis in original. Another study, of multinational pharmaceutical firms operating in Nigeria, found that most such firms 'especially through their affiliates and subsidiaries, overpromote their drugs for extra indications and with mention of fewer hazards', in comparison with information provided in the United States. See Osifo (1983) p. 8.
36. Greenhalgh (1987).
37. Krishnaswamy et al. (1985) p. 367.
38. For a detailed review of these and other factors that contribute to patient compliance in the developed countries, see Blum (1981).
39. According to a survey released in 1995 by the American Lung Association. See *USA Today*, 13 October 1995.
40. For the case of Sri Lanka, see Wolffers (1987); the Philippines, Hardon (1987); India, Greenhalgh (1987) and Krishnaswamy et al. (1985); Ethiopia, Sekhar et al. (1981).
41. Hogerzeil (1986) p. 436.
42. Some antibiotics, such as amoxicillin, for example, deteriorate without refrigeration.
43. This approach is used by Bennett et al. (1990).
44. Ibid.
45. Ibid.
46. The Sen approach is applicable in principle not just to medicinal drugs but to all products, which generally contribute to a wide range of human functionings, social and psychological, as well as physical. Food, for example, as Sen himself points out, not only contributes to one's nutritional functionings but also plays an important role in a person's social relationships. But whereas the former type of functionings can be objectively measured, many of the latter types of functionings are usually difficult to measure and are thus less amenable to analysing the effect of consumption on welfare in the same empirical way as we have done.

For many products, therefore, it seems to us that future research should focus on measurement issues related to the subjective connection between consumption and psychological and sociological functionings.

47. Cowan and Heap (1993).
48. Bennett et al. (1990).
49. Jelliffe and Jelliffe (1978).

References

K. Bennett, P. Tugwell, D. Sackett and B. Haynes (1990). 'Relative Risks, Benefits, and Costs of Intervention', in K. Warren and A. Mahmoud (eds), *Tropical and Geographical Medicine*, 2nd edn, Maidenhead: McGraw-Hill.

R. Blum (1981). 'Factors Affecting Individual Use of Medicines', in R. Blum et al. (eds), *Pharmaceuticals and Health Policy*, New York: Holmes & Meier.

G. A. Cohen (1993). 'Equality of What? On Welfare, Goods and Capabilities' in M. Nussbaum and A. Sen (eds), *The Quality of Life*, Oxford: Clarendon Press.

G. Cowan and B. Heap (1993). *Clinical Tropical Medicine*, London: Chapman & Hall.

T. Greenhalgh (1987). 'Drug Prescription and Self-Medication in India: An Exploratory Survey', *Social Science Medicine*, vol. 25, no. 3.

A. Hardon (1987). 'The Use of Modern Pharmaceuticals in a Filipino Village: Doctor's Prescription and Self Medication', *Social Science Medicine*, vol. 25, no. 3.

G. Helleiner (1975). 'The Role of Multinational Corporations in the LDC's Trade in Technology', *World Development*, vol. 3, no. 4.

H. Hogerzeil (1986). 'Use of Essential Drugs in Rural Ghana', *International Journal of Health Services*, vol. 16, no. 3.

J. James and F. Stewart (1981). 'New Products: A Discussion of the Welfare Effects of the Introduction of New Products in Developing Countries', *Oxford Economic Papers*.

D. Jelliffe and E. Jelliffe (1978). *Human Milk in the Modern World*, Oxford University Press.

R. Jenkins (1988). 'Transnational Corporations and Third World Consumption: Implications of Competitive Strategies', *World Development*, vol. 16, no. 11.

K. Krishnaswamy, B. Kumar and G. Radhaiah (1985). 'A Drug Survey – Precepts and Practices', *European Journal of Clinical Pharmacology*, vol. 29.

E. Martin (1978). *Hazards of Medication*, 2nd edn. Philadelphia: J. B. Lippincott.

M. Muller (1982). *The Health of Nations*, London: Faber & Faber.

N. Osifo (1983). 'Overpromotion of Drugs in International Product Package Inserts', *Tropical Doctor*, January.

H. Rang, M. Dale and J. Ritter (1995). *Pharmacology*, Edinburgh: Churchill Livingstone.

A. Schapira (1990). 'The Resistance of Falciparum Malaria in Africa to 4-Aminoquinolines and Antifolates', *Scandinavian Journal of Infectious Diseases*, Supplement 75.

K. Schmidt (1995). 'Inbred Parasites May Spur Resistance', *Science*, vol. 269, 22 September.

C. Sekhar, R. Raina and G. Pillai (1981). 'Some Aspects of Drug Use in Ethiopia', *Tropical Doctor*, July.

A. Sen (1985) *Commodities and Capabilities*, Amsterdam: North Holland.

M. Silverman and P. Lee (1974). *Pills, Profits and Politics*, University of California Press.

M. Silverman, P. Lee and M. Lydecker (1982). *Prescriptions for Death: the Drugging of the Third World*, University of California Press.

R. Snyder, H. Mofenson and J. Greensher (1973). 'Toxicity from Lomotil', *Clinical Pediatrics*, January.

F. Stewart (1977). *Technology and Underdevelopment*, London: Macmillan.

D. Weatherall (1990). 'Common Genetic Disorders in the Tropics', in K. Warren and A. Mahmoud (eds), *Tropical and Geographical Medicine*, 2nd edn, Maidenhead: McGraw-Hill.

I. Wickramasekera (1985). 'A Conditional Response Model of the Placebo Effect: Predictions from the Model' in L. White, B. Tursky and G. Schwartz (eds). *Placebo: Theory, Research and Mechanisms*, New York: Guildford Press.

I. Wolffers (1987). 'Drug Information and Sale Practices in Some Pharmacies of Colombo, Sri Lanka', *Social Science Medicine*, vol. 25, no. 3.

Part Two
Macro Aspects and Applications

5 The Employment Effects of an Income Redistribution in Developing Countries

The 1970 ILO Mission Report to Colombia first proposed the notion that since the poor tend to consume a more labour-intensive basket of commodities than the rich, a redistribution of income in their favour will tend to increase employment. This proposal gave rise to a host of attempts in the 1970s to demonstrate its empirical validity in a number of developing countries. The results however provided only a very limited degree of support to the ILO hypothesis. Indeed, by 1978 'the clear conclusion of the general equilibrium macroeconomic models' was that 'income redistribution would have only a tiny effect on employment'.[1]

Our goal here is to challenge this well-established conclusion, which, we believe, is based on a methodology that is inadequate in several important respects. To redress these methodological deficiencies, we advocate a framework that employs a highly disaggregated social accounting matrix (SAM) and on the basis of the empirical results that are derived therefrom, we suggest that the received view of the relationship between income redistribution and employment may contain a systematic understatement of the true effect.

A CRITIQUE OF THE RECEIVED METHODOLOGY

The methodology that was commonly employed to estimate the employment effects of an income redistribution during the 1970s has been lucidly summarized by Morawetz in the following terms.

> Models designed to examine the impact of income redistribution on aggregate employment generally begin by assuming a specific pattern of income redistribution, based on targets for the lowest group, on social objectives, or on taxation possibilities. The impact of the redistribution on savings and total investment is calculated from aggregate consumption functions for each group, plus assumptions as to external capital flows. The total consumption of each income

> group is then broken down into its component parts, using Engel curves or some other kind of demand function, and the new levels of total consumption for each commodity are fed into an input-output system in order to determine direct and indirect changes in production, imports, employment, and income distribution. Finally, on the basis of these changes, the more complete models iterate back through the system. Production, imports, and employment can now be compared to projections of their magnitudes assuming that there was no initial income redistribution.[2]

Though one can point to a number of weaknesses in this approach (as does Morawetz himself), we focus on two areas in particular, which, in combination, are shown to give rise to the need for an alternative approach. The first area is concerned with the level of aggregation at which the research during the 1970s was typically conducted.

Why the level of aggregation matters

Aggregation takes many different forms in economics: industries, sectors, households and firms, for example, are commonly added together for various purposes, such as the construction of an aggregated production or consumption function. It is intuitively clear that aggregation is not likely to matter much if the units being added are alike in relevant respects – if, for instance, 'the firms that are being added are very much alike – in their output (they may all produce one inch pins, for example), their inputs and their techniques of production'.[3] But when instead the units to be added together are highly heterogeneous there is a correspondingly greater degree of scope for some form of aggregation bias.

In the context of the highly dualistic structures of production and consumption that are characteristic of most developing countries, it is easy enough to argue that heterogeneity rather than homogeneity will be the more prevalent. And it is just as easy on these grounds to be critical of the methodology employed in the 1970s, which relied on models that were not only highly aggregated in terms of the factors of production, but also in terms of output which typically was classified according to two digits of the Standard Industrial Trade Classification (SITC) – thus effectively assuming that products within the broad categories, such as clothing and transport, were highly similar.[4] What is more difficult to contend, however, is that the heterogeneity which this methodology overlooks gives rise to an aggregation bias of a systematic kind (as opposed, for example, to errors in offsetting directions). For what this requires is some systematic pattern of association between the (heterogeneous) variables that would not be apparent at an aggregative level of analysis.[5]

Our view that some such association does indeed exist has as its point of departure the recognition that developing countries are characterized by the coexistence of forms of production and consumption which originate in very different historical periods.[6] For one thing, much of the production structure of these economies contains modes of production which belong to the period of 'pre-capitalist economic formations'. In this period, as Marx put it, 'the greater part of the products are produced for the satisfaction of the immediate needs of the community not as commodities',[7] in contrast to the later capitalist mode of production which turns 'the commodities formerly produced as immediate use-values into exchange values'.[8] At a later stage there may be simple as opposed to expanded commodity production. In such a situation the scale of operation is small and the technology used is mostly indigenous.

The origin of the historical relationship between products, technologies and incomes can thus be said to reside in communities where the mode of production exists essentially to provide for the subsistence needs of the community. That these earlier modes of production continue to survive in developing countries largely reflects the fact that (especially in Africa and parts of Asia) vast numbers still live at subsistence levels of income and thus demand the type of commodity which is appropriate to the earlier historical period. For the most part, following Lancaster's theory of demand that views commodities in terms of the characteristics they embody,[9] this means commodities containing a high proportion of essential characteristics, that are produced by self-employed and other small-scale producers, in and for impoverished groups of households typically in rural areas.

The organized or formal sector of the developing economy, on the other hand, operates mostly on the basis of imported processes of production developed in and for the incomes and product requirements of advanced economies. These requirements are generally such that product designs come to acquire an increasingly high proportion of 'high-income' (including labour- and time-saving) characteristics. Partly as a consequence, the techniques required to produce them become increasingly capital-intensive and large scale. One important dimension of product quality, for instance, is based on the need for 'close tolerances' and this is often very difficult to achieve without relatively sophisticated capital equipment.[10] What evolves consequently is a generally close relationship over time between products (defined as above in terms of embodied characteristics) and the nature of production processes with both being related to changes in the level of income per head (as already noted in Part I above).

Broadly speaking, therefore, the relationships between products, processes and incomes that can currently be observed in a developing country – that is, cross-sectionally – are in fact the reflection of what has occurred historically, ranging all the way from pre-capitalistic economic formations to the most modern products and processes, that, frequently, are associated with the activities of multinationals from the advanced countries. More specifically, what is suggested by this historical approach is a more or less systematic association between product characteristics, the factor intensity of technologies and household incomes that would not be discernible in the macro models of the 1970s and which would cause these models to understate the extent to which an income redistribution in favour of the poor is likely to promote employment.

Underlying this historically based tendency, as we see it, is the important notion of a *technological system*. What this notion seeks essentially to convey is the idea that: 'There are technological linkages between different parts of the system which mean that *much of technology comes as a package*.'[11] This package includes the relationships not only between products, techniques and incomes that were suggested above, but also between these and other variables such as infrastructural requirements, the nature and quality of inputs, the legal and administrative system, culture and so on.

In highly traditional communities, for example, the technology system has a strong geographical component; production and consumption activities, that is to say, tend to be closely related in a particular (and often isolated) rural locality, which relies relatively heavily on unskilled labour and self-employment as a mode of production and which exhibits relatively few links to external technological systems (or, put more technically, which has minimal leakages out of the system, as occurs, for example, through imported inputs of various kinds). Such a system, moreover, is likely to use (local) labour and other inputs relatively intensively: that is, with a high ratio of inputs per unit of output (reflecting, among other things, the lack of modern, productive technologies and advanced technological capabilities).

In the modern sector of the typical developing country, in contrast, the technological system tends closely to approximate what is found in the developed world: in particular, techniques that 'produce high-income products, require high levels of investible resources per employee, high levels of education and skills, be of a large scale and require sophisticated management techniques, be associated with high levels of labour productivity, and be linked, through inputs and outputs, with the rest of the advanced technology system'.[12] If this system is to be used efficiently in the developing country setting, it needs, as Stewart[13] has emphasized, to be reproduced and this in turn often gives rise to a heavy dependence on imported inputs of labour and other resources.

In contrasting traditional and advanced technological systems in this way, it is not our intention to suggest that these (and other intermediate) systems never interact or that they can always be so clearly distinguished from one another. What we wish to emphasize is rather that much of technology comes as a package and that the relationships between the components of this package need to be understood from a historical perspective.

In fact, as described in Chapters 2 and 3 above, there is already quite a lot of evidence to support (some aspects of) this way of looking at technological issues in developing countries. A number of micro case studies, for example, reveal the relationship between consumption patterns of different income groups and the factor intensity of production that one would expect on historical grounds, namely, for the 'poor' to purchase relatively 'low-income' and labour-intensive products and for the 'rich' to acquire 'high-income' and capital-intensive commodities.[14] Indeed, in some of these studies something like a continuum seemed to exist – with rising incomes consumers could be viewed as passing through the product and process characteristics of the earliest pre-capitalist formations to those of successively later periods.[15] In other, less 'pure', cases the same relationships, though less clear, were nevertheless discernible.

Such studies as these however cast only limited doubt on the received view of the employment effect of an income redistribution in favour of the poor.[16] The reason is that whereas this view is based on macro – or economy-wide – models, the studies we have just cited deal only with the direct (or partial equilibrium) effects of alternative technological choices. And especially since the direct effects are likely in general to comprise a relatively small proportion of the total (that is, direct and indirect) effect, they cannot readily be used as a proxy for the latter. What is needed instead therefore, if the received view is to be challenged, is a framework which incorporates both direct and indirect employment effects and it is to the choice of this framework that we now turn.

Input–output versus a social accounting matrix (SAM) approach

As already noted, the economy-wide models of the 1970s were based on an input–output framework for estimating the employment effect of an income redistribution. While there is little doubt that this framework provides some indication of the total effect, by no means can it be said to capture all the relevant economic relationships.[17]

In particular, whereas the input–output methodology shows how a change in final demand affects output in all other supplying sectors, it does not take into account

> the further consequences of increased sectoral outputs on the employment of labour and other factors of production which generates a stream of income received by these factors. In turn, this value added ends up – assuming no supply constraints (i.e. excess capacity) – as income to the various socioeconomic households and other institutions in the economy allowing the latter to spend more on consumption of the various commodities produced by the production activities thereby generating additional production. By excluding the link between increased output, the factorial and household income distribution and increased consumption the I/O approach short circuits the circular flow in the economy and thereby accounts for only a part of the indirect production in any of the production activities.[18]

Fortunately, an alternative framework, known as a social accounting matrix, is available which redresses these deficiencies of the input–output approach.[19] How this is achieved can be briefly illustrated with the aid of Table 5.1, which shows a simplified schematic social accounting matrix. As with any SAM, it consists of as many rows as columns. Expenditures are recorded by the latter and receipts by the former. With these conventions a SAM can be used as an economy wide accounting framework. As such a SAM is a consistent set of accounts since row totals (receipts) must equal column totals (expenditures). At the same time a second use of a SAM is in economic modelling. The descriptive accounting framework can be transformed to a basis for modelling by selecting some accounts as endogenous so that these can be determined within the model by the exogenous accounts or variables. Table 5.1 assumes that three accounts are

Table 5.1 Simplified schematic social accounting matrix

		Expenditures				
		Endogenous accounts			*Exogenous*	
Receipts		*Factors*	*Households*	*Production activities*	*Sum of other accounts*	*Totals*
		1	2	3	4	5
Endogenous accounts						
Factors	1	0	0	T_{13}	x_1	y_1
Households	2	T_{21}	T_{22}	0	x_2	y_2
Production activities	3	0	T_{32}	T_{33}	x_3	y_3
Exogenous accounts						
Sum of other accounts	4	$1'_1$	$1'_2$	$1'_3$	t	y_x
Totals	5	y'_1	y'_2	y'_3	y_x	

Source: Defourney and Thorbecke (1984).

endogenously determined, namely, factors, institutions (households and companies) and production activities, while all other accounts are assumed to be exogenous (that is, governments, capital and the rest of the world). 'Thus, for example, T_{13} is the matrix which allocates the value added generated by the various production activities into income accruing to the various factors of production and T_{33} shows the intermediate input requirements (i.e. the input–output transactions), while T_{32} reflects the expenditure pattern of the various institutions including the different household groups for the commodities (production activities) which they consume.'[20] As such, therefore, the input–output (or backward) linkages comprise only part of a much wider set of macro-economic relationships in the SAM.

Under conditions of excess capacity that permit prices to remain constant, the framework shown in Table 5.1 can be employed to estimate the effects of exogenous changes (such as in government expenditure or exports) on the entire system. The logic underlying any such exercise is that exogenous changes (the x's in Table 5.1) determine via the SAM matrix, the incomes of factors (the vector y_1); of households and companies (y_2); and of production activities (y_3). Conversion of the transaction matrix produces the corresponding matrix of average expenditure propensities. These are obtainable simply by division of a particular element in any of the endogenous accounts by the total expenditure for the column account in which the element appears.

More formally, if A_n is the square matrix of average expenditure propensities for the endogenous accounts in Table 5.1, endogenous incomes y_n (i.e. factor incomes y_1; institutional incomes y_2; and production activities incomes y_3 in that table) are equal to $A_n y_n + x$. Therefore,

$$\begin{aligned} y_n &= (I - A_n)^{-1}x \\ &= M_a x \end{aligned} \tag{1}$$

Thus, endogenous incomes can be derived by multiplying injection x by a multiplier matrix M_a, which is referred to as an accounting multiplier matrix. One limitation of this matrix, as derived in equation (1) is its implication of unitary expenditure elasticities (that is, that the average expenditure propensities in A_n are assumed to apply to any injection). It is more realistic to specify a matrix of marginal expenditure propensities (C_n below), which corresponds to the observed income and expenditure elasticities of the various agents, assuming that prices remain fixed. Formally, expressing the changes in incomes dy resulting from changes in injection dx, one derives:

$$\begin{aligned} dy_n &= C_n dy_n + dx \\ &= (I - C_n)^{-1}dx = Mcdx \end{aligned} \tag{2}$$

Mc is referred to as a fixed price multiplier matrix and the significant role that it plays in our empirical analysis will become apparent below. At this point, however, it bears emphasizing that – as the term 'fixed price multipliers' indicates – the initial redistribution leaves prices unaffected under the condition of excess capacity in productive activities. A more general approach which allows prices to be flexible is to be found in the literature on computable general equilibrium (CGE) models. Such models, however, have also been criticized for arbitrary choices of functional forms and parameter values. The approach taken in this paper is intended to be a first approximation, contrasting the general equilibrium (or disequilibrium) approach with the partial (equilibrium) methodology we criticize.

SAM-TECH Indonesia as a disaggregated framework for estimating the employment effect of an income redistribution

So far, in criticizing the received methodology, we have emphasized that estimation of the employment effect of an income redistribution needs to be conducted on a disaggregated basis and that it should incorporate a wider variety of macro-economic relationships than are contained in an input–output approach. Both these requirements are met in a modified SAM framework for Indonesia that Khan and Thorbecke[21] refer to as SAM-TECH.

What is novel about SAM-TECH is its attempt to disaggregate technologies and products in six sectors which exhibit technological dualism-sectors, that is, in which production takes place both in traditional, highly labour-intensive units as well as in more modern, capital-intensive enterprises. The distinction between the traditional and modern methods of production in each of these sectors (hand-pounded vs. milled rice; farm vs. plant processed tea; dried and salted vs. canned fish; brown vs. refined sugar; canning and preserving of fruits and vegetables in small vs. medium and large firms; and clove vs. white cigarettes), meant that in total 12 new production activities were grafted on to the 1975 SAM for Indonesia, which was constructed by the Central Bureau of Statistics. (Together these dualistic activities accounted for some 50 per cent of manufacturing sector output in 1975.)[22]

Apart from the specific (and substantial) attempt at disaggregation that this represents, it is important to emphasize that the 1975 SAM for Indonesia is *itself* very detailed. In particular, it contains 78 columns and rows, which include 16 different labour groups (classified by sector of employment, paid vs. unpaid and rural vs. urban) and ten socio-economic household groups. In short, therefore, SAM-TECH provides an unusually comprehensive framework for assessing the employment effect of an income redistribution and for analysing the role that disaggregation plays in this relationship. We examine next how various redistributive schemes were simulated within this framework and we report on the results that were obtained.

THE SIMULATIONS: METHODOLOGY AND RESULTS

The procedure adopted can be described in terms of the following five stages.

1. From population census and household budget survey data, we were able to obtain figures for the total number of households and total household income.[23] This enabled us to derive the income of each of the ten household groups classified in SAM-TECH.

2. We then devised three different redistribution schemes to transfer income from richer to poorer households, both rural and urban. The first scheme transfers 100,000 rupees from farm size 2 (representing farms owning between 0.51 and 1.00 Ha) to farm size 1 (representing farmers owning up to 0.5 Ha). The second scheme transfers 100,000 rupees from 'rural higher' households to 'rural lower' households. The last scheme transfers 100,000 rupees from 'urban middle' and 'urban higher' to 'urban lower' households.[24]
3. Using estimated consumption functions, we calculated the changes in expenditure of the affected households. Ideally, we would have wished to include all six of the dualistic sectors from SAM-TECH in these estimations, but the consumer expenditure survey data were compatible with the production-side classifications in only one of these sectors, namely, sugar. For this sector, separate consumption functions were estimated for brown sugar (the traditional unrefined product), white sugar (the modern refined product), and total sugar. The three functions estimated on the basis of weighted ordinary least squares [OLS] (with weights given by the number of households in each group) are as follows.

$$\text{Brown sugar } b = -\underset{(6029.55)}{43419} + \underset{(460.96)}{3569.69} \times \log(y) \quad \bar{R}^2 = 0.87$$

$$\text{White sugar } s = -\underset{(8589.02)}{110470} + \underset{(656.63)}{9337.21} \times \log(y) \quad \bar{R}^2 = 0.96$$

$$\text{Total sugar } t = -\underset{(10202.02)}{153889} + \underset{(779.94)}{12907} \times \log(y) \quad \bar{R}^2 = 0.97$$

These estimated non-linear consumption functions allow us to integrate econometric modelling with fixed price multiplier modelling based on the SAM. The partial equilibrium expenditure effects of an income redistribution are captured initially through the above functions. Later the fixed price multipliers are used to capture the general equilibrium impact on employment. Thus, application of these functions to the household groups affected by the three redistributive schemes, yields the changes in expenditure shown in Table 5.2. In each case the change in expenditure on total sugar is made up (in roughly the same proportions) of expenditure changes on the two different types of sugar (and it is worth noting at this point that the respective shares of brown and white sugar in 1975 were of the order of 25 and 75 per cent).[25]

4. The changes in expenditure of the affected households that are shown in Table 5.2 were multiplied by the fixed price multipliers (FPM) derived from the Indonesian SAM, for the sixteen labour factors that appear there (See Tables 5.3, 5.4 and 5.5 below).

Table 5.2 Change in expenditure (using the semi-log function)

	Sugar Rs	*Brown sugar Rs*	*White sugar Rs*
Transfer of Rs 100,000 from FS 2 to FS 1	6461.438	1787.04	4674.347
Transfer of Rs 100,000 from RH to RL	5114.33	1414.47	3699.821
Transfer of Rs 100,000 from UM to UL	309.808	85.67	224.1007

Table 5.3 Changes in employment – redistribution scheme 1

Labour Categories	*Aggregated SAM*	*Disaggregated SAM*	
	Sugar	*Brown sugar*	*White sugar*
Agric. paid rural	1.033	3.131	4.865
Agric. paid urban	1.098	0.090	0.142
Agric. unpaid rural	1.227	5.471	9.322
Agric. unpaid urban	1.227	0.179	0.313
Prod. paid rural	0.839	1.669	3.589
Prod. paid urban	1.292	0.461	1.700
Prod. unpaid rural	0.904	1.925	3.021
Prod. unpaid urban	1.292	0.189	0.525
Clerical paid rural	0.839	0.720	1.465
Clerical paid urban	1.227	0.005	1.386
Clerical unpaid rural	0.904	1.891	3.609
Clerical unpaid urban	1.227	0.698	1.389
Prof. paid rural	0.904	0.230	0.468
Prof. paid urban	1.227	0.183	0.428
Prof. unpaid rural	0.904	0.123	0.277
Prof. unpaid urban	1.292	0.089	0.200
Total	17.501	17.053	32.709

5. The resulting figures were divided by the average wages/salaries of the respective labour factors (also derived from the SAM) to estimate the effects on employment. These estimates are shown in Tables 5.3, 5.4 and 5.5. for the three different redistributive schemes.

In each of these tables, income redistribution in favour of the poor does have a favourable (direct plus indirect) effect on employment, at both aggregated and disaggregated levels of analysis. In each case, however, the overall effect is more pronounced at the latter level – when, that is to say, the combination of employment increases in the brown and white sugar subsectors, exceeds, by a factor of about three, the estimate for total sugar.

In interpreting these results, which confirm our main hypothesis that aggregation tends to understate the employment effects of an income redistribution, it is essential to recognize that the overall understatement shown in Tables 5.3, 5.4 and 5.5. does not occur *uniformly* across all 16 labour factors. On the contrary, there is a substantial degree of variation across these factors. On the one hand, as is evident from Table 5.3 for example, in seven of the labour categories (agricultural

Table 5.4 Changes in employment – redistribution scheme 2

Labour categories	*Aggregated SAM*	*Disaggregated SAM*	
	Sugar	*Brown sugar*	*White sugar*
Agric. paid rural	0.818	2.478	8.725
Agric. paid urban	0.869	0.071	0.112
Agric. unpaid rural	0.971	4.330	7.378
Agric. unpaid urban	0.971	0.142	0.248
Prod. paid rural	0.664	1.321	2.841
Prod. paid urban	1.022	0.365	1.345
Prod. unpaid rural	0.716	1.523	2.391
Prod. unpaid urban	1.022	0.150	0.416
Clerical paid rural	0.664	0.570	1.160
Clerical paid urban	0.971	0.434	1.097
Clerical unpaid rural	0.716	1.497	2.856
Clerical unpaid urban	1.022	0.552	1.100
Prof. paid rural	0.716	0.182	0.370
Prof. paid rural	0.971	0.144	0.339
Prof. unpaid rural	0.716	0.098	2.417
Prof. unpaid urban	1.022	0.070	0.158
Total	13.851	13.927	32.953

Table 5.5 Changes in employment – redistribution scheme 3

Labour categories	*Aggregated SAM*	*Disaggregated SAM*	
	Sugar	*Brown sugar*	*White sugar*
Agric. paid rural	0.049	0.150	0.233
Agric. paid urban	0.052	0.004	0.006
Agric. unpaid rural	0.058	0.262	0.446
Agric. unpaid urban	0.058	0.008	0.015
Prod. paid rural	0.040	0.080	0.172
Prod. paid urban	0.061	0.022	0.081
Prod. unpaid rural	0.043	0.092	0.144
Prod. unpaid urban	0.061	0.009	0.025
Clerical paid rural	0.040	0.034	0.070
Clerical paid urban	0.058	0.026	0.066
Clerical unpaid rural	0.043	0.090	0.173
Clerical unpaid urban	0.061	0.033	0.066
Prof. paid rural	0.043	0.011	0.022
Prof. paid urban	0.058	0.008	0.020
Prof. unpaid rural	0.043	0.005	0.013
Prof. unpaid urban	0.061	0.004	0.009
Total	0.829	0.838	1.561

paid urban, agricultural unpaid urban, production unpaid urban and the last four professional categories), the aggregated estimate (for total sugar) *exceeds* the sum of the two disaggregated estimates (so that there is overstatement rather than understatement). On the other hand, much of the overall understatement is concentrated in four labour categories: agricultural paid rural, agricultural unpaid rural, production paid rural and production unpaid rural. These, unlike most of the seven cases of overstatement that have just been mentioned, are labour categories drawn from the rural (rather than the urban) areas and they contain a relatively high proportion of unskilled (and low-income) workers.[26]

Underlying these contrasting patterns are in part the relevant fixed price multipliers shown in Table 5.6 (as well of course as the particular patterns of consumption that are induced by the income redistribution).

These provide, in the first place, part of the reason why understatement of the employment increase at the aggregate level is concentrated in the four rural labour categories noted above in relation to Table 5.3. In fact, the fixed price multipliers for brown sugar are highest in precisely these categories and much the same is true for white sugar as well. Evidently, it is in these labour categories that the aggregate SAM most sharply underestimates the total linkages. Correspondingly, when these groups are the beneficiaries of an income redistribution (as in our first two simulations), the employment effect will tend to be large in comparison to cases where (as in our third simulation) it is urban households who are favourably affected.

The disaggregated multipliers in Table 5.6 also lend some degree of support to our view that historically determined relationships between products, technologies and incomes in distinct 'systems', give rise to aggregation bias of the sort found here. For what that table reveals are uniformly higher fixed price multipliers for brown than for white sugar. In addition, the former are especially high among the rural labour groups whose consumption patterns, we suggested earlier, would be relatively intensive in labour and other local inputs and which would exhibit relatively few outside leakages.

One can in fact get a rather better sense of the differential linkages that underlie these dissagregated fixed price multipliers, by means of a relatively new technique known as 'structural path analysis', which, in effect, decomposes the multipliers into a series of constituent influences.

Table 5.6 Disaggregated fixed price multipliers

Labour categories	*Aggregated Sugar*	*Disaggregated*	
		Brown sugar	*White sugar*
Agric. paid rural	0.012	0.133	0.079
Agric. paid urban	0.016	0.005	0.003
Agric. unpaid rural	0.013	0.218	0.142
Agric. unpaid urban	0.016	0.009	0.006
Prod. paid rural	0.016	0.118	0.097
Prod. paid urban	0.038	0.049	0.069
Prod. unpaid rural	0.014	0.110	0.066
Prod. unpaid urban	0.032	0.017	0.018
Clerical paid rural	0.014	0.045	0.035
Clerical paid urban	0.038	0.061	0.059
Clerical unpaid rural	0.011	0.085	0.062
Clerical unpaid urban	0.024	0.046	0.035
Prof. paid rural	0.049	0.045	0.035
Prof. paid urban	0.106	0.057	0.051
Prof. unpaid rural	0.014	0.007	0.006
Prof. unpaid urban	0.084	0.021	0.018

Indeed, Khan and Thorbecke show that when this technique is applied to SAM-TECH it is capable of capturing 'the entire network of paths through which the impact of a change in demand for the output in any production activity is transmitted throughout the Indonesian economy'.[27] Sugar is one of the sectors used by them to illustrate how these paths actually operate and, more specifically, how the effects of a change in the output of brown and white sugar ultimately get translated into higher incomes for the household group headed by small farmers. What Khan and Thorbecke conclude from this exercise is that although both types of sugar 'affect the economic system through the same set of paths', the major difference between them consisted of the 'considerably smaller backward linkages to sugar cane... of the modern technological alternative (refined sugar) compared to the traditional one (brown sugar) which is made out of palm sugar'.[28]

CONCLUSIONS

During the 1970s a host of studies attempted to examine the employment effects of an income redistribution in favour of the poor in developing countries. These studies however provided little or no support for the proposition that since the poor tend to consume a more labour-intensive basket of commodities than the rich, a redistribution in their favour will tend to increase employment.

We have argued that the methodology underlying these results is flawed in two (related) ways, the first being the use of a level of aggregation that disguises a historically based pattern of associations between product characteristics, technologies and incomes. Neglect of these relationships – which tend to manifest themselves in terms of distinct 'technological systems' – imparts a tendency for the aggregative approach to underestimate the true employment effects of an income redistribution. In addition, based as they generally are on an input–output approach, the models used during the 1970s neglect a variety of relevant macro-economic relationships. In particular, the input–output approach excludes the link between increased output, the factorial and household income distribution and increased consumption.

To redress these methodological deficiencies, we advocated an alternative analytical framework based on a disaggregated SAM for Indonesia. This framework is highly disaggregated not only in terms of products and technologies for six dualistic manufacturing sectors, but also in terms of labour factors and households. On the basis of this framework a number of income redistributions from rich to poor was simulated, using consumption functions estimated for one dualistic sector, sugar, at both aggregative and disaggregative levels of analysis. Though employment was found to increase at both levels, the effect is far more pronounced in the disaggregative case. Analysis of the sources of this discrepancy suggested that they were partly the result of a tendency for aggregation to most heavily understate linkage effects among relatively unskilled rural labour categories. To at least some extent this tendency seems to be explicable with reference to our theoretical framework which emphasizes the historical origins of the highly dualistic patterns of production and consumption in the manufacturing sector of most developing countries.

Notes

1. White (1978), p. 46. Morawetz (1974) comes to the same conclusion.
2. Morawetz (1974) p. 504.
3. Walters (1968) p. 308.
4. Morawetz (1974).
5. Some writers on this topic have explicitly questioned whether any such association exists. Cline (1975), for example, argues that there are no theoretical reasons to expect systematic factor combination differences between basic and luxury goods at a more disaggregated level. More recently, Cooper (1990) has expressed a similar degree of scepticism.
6. This argument is based on Chapter 2 above.
7. Marx (1946) p. 377.
8. Marx (1964) p. 116.
9. Lancaster (1966).
10. Forsyth, McBain and Solomon (1982).
11. Stewart (1977), p. 6, emphasis added. The following discussion of a 'technology system' is based largely on this reference.
12. Stewart (1981) p. 94.
13. Stewart (1977).
14. See van Ginneken and Baron (1984); James and Stewart (1981); Stewart (1977), chapters 9 and 10.
15. Chapter 2.
16. For the sugar processing industry in India, Chapter 3 above shows that a 50 per cent understatement of the employment effect of an income redistribution can be attributed to aggregation bias, i.e. the bias that results from failing to decompose the demand for sweeteners into its constituent parts.
17. This section very closely follows the text in Khan and Thorbecke (1988, 1989).

18. Khan and Thorbecke (1988), p. 2.
19. Already in the 1970s, several 'first generation' models had directly addressed the link between income distribution and employment within a general equilibrium framework. (Thorbecke and Sengupta 1972, and Pyatt et al. 1972). Both models were prepared under the auspices of the World Employment Programme of the ILO and they were both influential in the opening up of the SAM and CGE approaches. These two models are discussed in Pyatt and Round (1985).
20. Khan and Thorbecke (1989) p. 137.
21. Ibid.
22. Khan and Thorbecke (1988) p. 147.
23. Our consumption data are taken from the Indonesian 1980 Household Budget Survey (called SUSENAS), while the numbers of households are those reported in Keuning (1988).
24. In principle each of these redistributive schemes could be brought about by some combination of taxes from and subsidies to, the groups which the SAM identified. In practice, however, any such measures are likely to encounter not only administrative but also political problems with regard to the question of political will, which will influence the feasibility of redistribution even in the relatively mild changes that we have simulated. For a detailed discussion of these issues see Frank and Webb (1977).
25. Khan and Thorbecke (1988) p. 147.
26. Ibid.
27. Khan and Thorbecke (1989) p. 147.
28. Ibid, p. 155. On structural path analysis see also the following chapter.

References

W. Cline (1975), 'Distribution and Development: a Survey of Literature', *Journal of Development Economics*, vol. 1.

C. Cooper (1990), 'Notes on Income Distribution, Growth and Demand', in F. Stewart, H. Thomas and T. de Wilde (eds), *The Other Policy*, London: IT Publications.

J. Defourney and E. Thorbecke (1984), 'Structural Path Analysis and Multiplier Decomposition within a Social Accounting Matrix Framework', *Economic Journal*, vol. 94, March.

D. Forsyth, N. McBain and R. Solomon (1982), 'Technical Rigidity and Appropriate Technology in Less Developed Countries', in F. Stewart and J. James (eds), *The Economics of New Technology in Developing Countries*, London: Frances Pinter.

C. Frank and R. Webb (eds) (1977), *Income Distribution and Growth in the Less-Developed Countries*, Washington, DC: Brookings Institution.

W. van Ginneken and C. Baron (eds) (1984), *Appropriate Products, Employment and Technology*, London: Macmillan.

J. James and F. Stewart (1981), 'New Products: a Discussion of the Welfare Effects of the Introduction of New Products in Developing Countries', *Oxford Economic Papers*, vol. 33, no. 1.

S. J. Keuning (1988), 'Comparing a System of Socio-Economic Accounts for Indonesia for 1975 and 1980: Documentation of the Compilation of the Data, Part 1', Statistical Analysis Capability Programme, Working Paper Series No. 1, The Hague: Institute of Social Studies.

H. Khan and E. Thorbecke (1989), 'Macroeconomic Effects of Technology Choice: Multiplier and Structural Path Analysis within a SAM Framework', *Journal of Policy Modelling*, vol. 11, no. 1.

H. Khan and E. Thorbecke (1988), *Macroeconomic Effects and Diffusion of Alternative Technologies Within a Social Accounting Matrix Framework*, Aldershot: Gower.

K. Lancaster (1966), 'A New Approach to Consumer Theory', *Journal of Political Economy*, vol. LXXIV, April.

K. Marx (1964) 'Pre-capitalist Economic Formations', in E. J. Hobsbawm (ed.), *Pre-Capitalist Economic Formations*, London: Lawrence & Wishart.

K. Marx (1946), *Capital*, London: Allen & Unwin.

D. Morawetz (1974), 'Employment Implications of Industrialisation in Developing Countries: a Survey', *Economic Journal*, vol. 84, September.

G. Pyatt, and J. Round (eds) (1985), *Social Accounting Matrices: A Basis for Planning* (Washington DC: World Bank).

G. Pyatt et al. (1972), *Employment and Income Policies for Iran*, Geneva: ILO.

F. Stewart (1981), 'International Technology Transfer: Issues and Policy Options', in P. Streeten and R. Jolly (eds), *Recent Issues in World Development*, Oxford: Pergamon.

F. Stewart (1977), *Technology and Underdevelopment*, London: Macmillan.

E. Thorbecke and J. Sengupta (1972), 'A Consistency Framework for Employment, Output and Income Distribution Projections Applied to Colombia', mimeo, Washington DC: World Bank Development Research Center.

A. A. Walters (1968), *An Introduction to Econometrics*, London: Macmillan.

L. White (1978), 'The Evidence on Appropriate Factor Proportions in Less Developed Countries: a Survey', *Economic Development and Cultural Change*, vol. 27.

6 Technology Choice and Income Distribution

Within the context of the agricultural sector one can point to numerous attempts to assess the macro-economic effects of alternative technologies on the distribution of income. One of the most influential early efforts, for example, was undertaken mainly in the Indian context by John Mellor,[1] who was seeking to redefine agriculture's role in development strategy and policy. In rejecting the then existing capital-intensive and industry-led development model and advocating instead a rural-led, employment-intensive path to development, Mellor was concerned, among other things, with the relationships between increased foodgrain production using alternative technologies and the growth rate of non-agricultural employment, non-agricultural sector capital–labour ratios and the course of per capita incomes of the labour force. More recently, Haggblade and Hazell[2] have used an input–output based model to examine how differences in the choice of agricultural technology and farm size influence the indirect (or multiplier) effects of increases in agricultural output. They show, for example, that: 'The input intensity, consumption profile of targeted farms and processing characteristics of the farm output all affect the size and composition of nonfarm spinoffs.'[3] The finding that consumption linkages generally dominate the total multiplier confirms Mellor's emphasis on the importance of this particular type of linkage effect.

In the area of non-agricultural technology and development, by contrast, the relationship between technological choice and income distribution has been largely neglected. To a degree, this reflects the fact that the literature is overwhelmingly micro-economic in character and is therefore able, by its very nature, to capture only the direct effects of alternative technologies on the distribution of income. The indirect or feedback effects, which, in practice, tend to be far more important, need to be estimated within a macro rather than a micro-economic framework. The problem, however, is that this is difficult to accomplish on the basis of most existing macro-economic models, because these do not usually differentiate between alternative technologies and products. Some way needs to be found therefore of incorporating micro-economic technological data into these models, which would then be capable of estimating the indirect as well as the direct effects of alternative technology choices on the distribution of income.

If we are substantially to improve our understanding of the topic, however, we cannot confine ourselves only to the empirical issue, for there is also a need to provide an analytical framework that can be used to interpret the findings from the thus improved methodology. This chapter, accordingly, has a dual objective, the first being to provide a general discussion of the differential effects on the distribution of income that can be expected from traditional as opposed to relatively modern technologies. The second section then shows how this framework can be used to understand the short-run distributional impact of alternative sugar processing technologies in Indonesia, as estimated within a modified social accounting matrix (SAM).

THE ANALYTICAL FRAMEWORK

Production techniques can be said to embody a range of different characteristics, such as the type and nature of the product, the organization of production, raw material and labour inputs and the scale and location of production.[4] In orthodox economic theory, there is nothing that would suggest any systematic association between these various characteristics. Rather, one would expect a more or less random association of characteristics across techniques of varying labour-intensity.

An alternative view, however, is that different technologies form part of distinct *technological systems* and that within each such system predictable relationships emerge between the variables just mentioned. The theoretical basis for such a view resides mainly in the Marxist notion of a mode of production, with its emphasis on the 'systemic' aspects of societies and especially the interrelations between technology, property rights and preferences in each type of society.[5] The mode of production, he suggested,

> is an internally balanced whole in which production relations, and especially the base on which they rest – the ownership of the means of production – are adjusted to the requirements of a given state of the development of social productive forces. We know that production relations cannot be arbitrary for given productive forces. A large industrial unit like an iron works or a locomotive factory, where a great number of people are employed using large and

numerous specialized machines and technical installations, cannot exist in conditions of simple commodity production, where the ownership of the means of production is scattered among a very large number of individuals. Such a production unit requires either capitalist or socialist production relations. [6]

If a historical perspective thus helps to make a connection between the way production is organized, the type of technology and the scale of production, it also helps to provide associations between other characteristics of a production technology that were mentioned at the beginning of this section, thus further contributing to the idea that technology should be viewed in terms of a system rather than as a vector of isolated characteristics. Stewart, for example, has shown how, at any point in time, techniques tend to be:

> developed against a background of a particular technology package[7] ... Any single technical innovation has to fit in with the rest of the system both in terms of the requirements it imposes for inputs, and in terms of the demand for the goods. A new technique must use inputs that are available, or can be made available, and must provide output which will fit into further production if it is an intermediate good, or into consumption patterns if it is a consumer good.[8] ... There are technological linkages between different parts of the system which mean that much of technology comes as a package, which cannot be separated and introduced bit-by-bit, but which goes together. [9]

In most developing countries, however, there is no one technological system; rather a range of systems can be identified with features that may often be highly disparate, reflecting the historical conditions at a particular point of time. At the one extreme, for example, are technological systems which represent the period of 'pre-capitalist' economic formations. In these societies, the technological relationships within the system have a strong geographical component. Production and consumption activities, that is to say, tend to be closely related in a particular geographical location: 'the greater part of the products are produced for the satisfaction of the immediate needs of the community not as commodities'. [10] This (typically rather isolated) locality would usually exhibit a heavy degree of reliance on unskilled labour and self-employment (or family labour) as the mode of production, and it would usually exhibit minimal links with external technological systems (put otherwise, there would be highly limited leakages out of the system, as occur, for example, when imports from other systems take place). Such a technological system, furthermore, tends to make intensive use of local labour and other inputs per unit of output (as one would expect in a system lacking modern technologies and advanced technological capabilities).

At the other extreme, the modern sector of most developing countries closely resembles the technological system that is found in developed countries. This means, among other things, production techniques that are associated with the following characteristics:

Table 6.1 Alternative technology systems: modes of influence on income distribution

Traditional technology system	*Modern technology system*	*Mode of influence on income distribution*
(a) Labour-intensive methods of production	Capital-intensive methods of production	Direct employment effect
(b) Non-capitalist mode of production	Capitalist mode of production	Share of income accruing to factors of production
(c) Dispersed small-scale production units	Production concentrated in small number of large-scale units	Dispersion of incomes across production units
(d) Relatively high labour–output ratio throughout production chain	Relatively low labour–output ratio throughout production chain	Indirect employment effect
(e) Relatively high non-labour (input) to output ratio production chain	Relatively low non-labour (input) to output ratio throughout production chain	Backward linkages
(f) Uses mainly inputs from own system	Uses mainly inputs from from own system	Backward linkages
(g) Relatively low (savings and import) leakages from the system	Relatively high (savings and import) leakages from the system	Linkages
(h) Consumes a relatively high proportion of its own products	Consumes a relatively high proportion of its own products	Consumption linkages

Table 6.2 Alternative technology systems: implications for the distribution of income

Traditional technology system	*Modern technology system*
(a) Generates high amount of direct employment	Generates low amount of direct employment
(b) Relatively high percentage of value added accrues to rural self-employed, household family members	Relatively high percentage of value added accrues to (local and foreign) companies as profits
(c) Wide geographical dispersion of income among small-scale units	Concentrated income generation among large-scale units
(d) Generates high amount of indirect employment through backward linkage	Generates low amount of indirect employment through backward linkage
(e) Generates relatively large backward linkage effects	Generates relatively small backward linkage system
(f) Linkages generated mainly within the traditional system	Linkages generated mainly within the modern system
(g) Linkages within the system relatively unaffected by leakages	Linkages within the system subject to relatively substantial leakages
(h) Consumption linkage to (traditional) products within the system	Consumption linkage to (modern) products within the system

high-income sophisticated products; high levels of investment per head; educated and skilled labour inputs; high levels of labour productivity; close links, via backward and forward linkages to the modern technological system (frequently via a heavy dependence on imported inputs from developed countries).

Table 6.1 juxtaposes these contrasting aspects of traditional and modern technological systems and it also shows the mechanisms through which these differences are likely to bear on the distribution of income. In some cases, for example, the differences bear on the direct and indirect employment effects that can be expected from traditional as opposed to modern technologies, while in other cases the differences bear on backward linkage effects or the dispersion of incomes across production units of varying sizes. Table 6.2 combines the information contained in each row of Table 6.1 into a set of eight propositions about the differential effects of traditional and modern technologies on the distribution of income.

THE SAM FRAMEWORK

The empirical analysis that is described below is based on a social accounting matrix (SAM), albeit one with certain distinctive features, and the presentation of the results relies heavily on the concept of fixed price multipliers. In order to understand the derivation of these fixed-price multipliers let us consider the following schematic SAM where the accounts have been divided into endogenous and exogenous categories for modelling purposes.[11] In concrete terms, for the Indonesian case to be described below, the technology-cum-productive activities, factors and households are endogenous, while the remainder of the domestic economy as well as the rest of the world is exogenous.

Looking at Table 6.3, which represents a SAM, we can see immediately that:

$$y = n + x \tag{1}$$

$$y = 1 + t \tag{2}$$

Table 6.3 Schematic representation of endogenous and exogenous accounts in a SAM

	Expenditures				
	Endogenous	*Sum*	*Exogenous*	*Sum*	*Totals*
Endogenous	T_{nn}	n	Injections T_{nx}	x	y_n
Exogenous	Leakages T_{xn}	1	Residual balances T_{xx}	t	y_x
Totals	y_n		y_x		

Now, if we divide the entries in the matrix T_{nn} by the corresponding total income (i.e. y_n), we can define a corresponding matrix of average expenditure propensities. Let us call this matrix A. We now have:

$$\mathrm{y} = \mathrm{n} + \mathrm{x} = \mathrm{Ay} + \mathrm{x} \tag{3}$$

$$y = (\mathrm{I} - \mathrm{A})^{-1}\mathrm{x} = \mathrm{Mx} \tag{3.1}$$

M has been called the matrix of the accounting multipliers, because these multipliers, when computed, can account for the results (e.g. income, consumption, etc.) obtained in the SAM without explaining the process that led to them. Let us now partition the matrix A in the following way:

$$A = \begin{bmatrix} 0 & 0 & A_{13} \\ A_{21} & A_{22} & 0 \\ 0 & A_{32} & A_{33} \end{bmatrix}$$

Given the accounts factors, household, and production activities, we now see that the income levels of these accounts (call them y_1, y_2, y_3, respectively) are determined as functions of the exogenous demand of all other accounts. In this respect, what we have is a reduced-form model that can be consistent with a number of structural forms. This is satisfactory as far as tracing the effects of a certain injection in the economy is concerned, or for prediction purposes when the structural coefficients are more or less unchanged.

One limitation of the accounting multiplier matrix M as derived in Equation 3.1 is that it implies unitary expenditure elasticities (the prevailing average expenditures propensities in A are assumed to apply to any incremental injection). A more realistic alternative is to specify a matrix of marginal expenditure propensities (C_n below) corresponding to the observed income and expenditure elasticities of the different agents, under the assumption that prices remain fixed. Expressing the changes in incomes (dy_n) resulting from changes in injections (dx), one obtains:

$$\begin{aligned} dy_n &= C_n dy_n + dx \\ &= (I - C_n)^{-1} dx = M_c dx \end{aligned} \tag{4}$$

M_c has been coined a fixed price multiplier matrix, and its advantage is that it allows any non-negative income and expenditure elasticities to be reflected in M_c. In particular, in exploring the macro-economic effects of exogenous changes in the output of different commodities on other macro-economic variables, it would be very unrealistic to assume that consumers reacted to any given proportional change in their incomes by increasing expenditures on the different commodities by exactly that same proportion (i.e. to assume that the income elasticities of demand of the various socio-economic household groups for the various commodities were all unitary). Since the income elasticity is the ratio of marginal expenditure propensity (MEP_i) to the average expenditure propensity (AEP_i) for any given good i, it follows that the marginal expenditure propensity can be readily obtained once the income elasticity and the average expenditure propensities are known:

$$\epsilon y_1 = \frac{MEP_i}{AEP_i} \tag{5}$$

$$MEP_i = \epsilon y_i AEP_i \tag{6}$$

$$\text{and} \quad \Sigma MEP_i = 1 \tag{7}$$

Thus, given the matrix A_{32} of average expenditure propensities, and the corresponding income elasticities of demand y_i, the corresponding marginal expenditure propensities matrix C_{32} could easily be derived.

Before moving on to the case study in the next section some special features of the above modelling procedure should be noted. As noted already this is a fixed-price multiplier model which is appropriate for the short (or perhaps intermediate) run but not for the long run when prices are flexible. However, in economies such as Indonesia the implications of short-run rigidities may be captured by the fixed-price approach.

A second, related aspect has to do with the value of the multipliers. At full employment the impact of an increase in demand will be reflected in prices, not output and employment. The operating assumption here is, therefore, that in many sectors there is excess capacity. Given the existence of open and disguised unemployment in Indonesia this is not an unreasonable assumption.

Finally, some remarks are in order for the kind of factor markets assumed in the model. For (short-run) wage rigidities to be an appropriate assumption labour market imperfections must be present. The data and the discussion published by the BPS (Indonesian Central Bureau of Statistics) indicate widespread wage-rigidities in both agricultural and non-agricultural sectors. The Indonesian SAM was prepared by using this database, thus the assumption of labour market imperfections would seem to be consistent with the institutional realities of Indonesia. It should also be noted that different households have different savings rates and capital markets are also imperfect.

Keeping the above features in mind the present model can be seen as a first approximation to the *short-run* effects of technology on income generation in Indonesia.

AN INDONESIAN CASE STUDY

As noted earlier, most macro-economic models do not differentiate between alternative technologies and products and they are therefore incapable of capturing any differential distributional effects of such technologies. One exception to this general tendency, however, is the modified SAM framework for Indonesia, that Khan and Thorbecke[12] refer to as SAM-TECH.

This framework is exceptional because it attempts to differentiate between technologies and products in six manufacturing sectors where a marked degree of technological dualism exists: sectors, in other words, where traditional and modern methods of production coexist. After distinguishing between these alternative methods of production in each of six manufacturing sectors (hand-pounded vs. milled rice; farm vs. plant processed tea; dried and salted vs. canned fish; brown vs. refined sugar; canning and preserving of fruits and vegetables in small vs. medium and large firms; and clove vs. white cigarettes), Khan and Thorbecke were able to graft 12 new production activities onto the 1975 SAM for Indonesia. SAM-TECH is also disaggregated in terms of labour and household groups, containing, respectively, 16 and 10 such categories. As such, it is particularly suitable for testing the hypotheses advanced in Table 6.2. This could in principle be done for each of the six dualistic sectors just mentioned, but in this chapter we have chosen to present the results for just one sector, sugar, in the substantial degree of detail that is necessary for an examination of the various hypotheses contained in Table 6.2. The appendix to this chapter summarizes the results for the other dualistic sectors.

Alternative sugar processing technologies and household income distribution

Table 6.4 shows the economy-wide income effects on the ten household groups, of a unitary injection of exogenous demand for sugar produced on the basis of both traditional and modern technologies. That is, in the form of fixed-price multipliers, the table shows the effects of a unitary injection of demand on the ten household groups after all direct and indirect effects have been taken into account within the SAM.

An increase of one rupiah's worth of (brown) sugar produced by traditional technology, for example, would ultimately yield 0.192 units of income to households labelled agricultural employees and 0.342

units of income to households represented by small farms. Similarly, an increase of refined sugar by the same amount would yield 0.110 and 0.205 to these two groups, respectively.

The results contained in Table 6.4 are broadly consistent with our analytical framework in that the traditional technology benefits rural rather than urban households while the converse is true of the modern sugar technology. However, we also need to know whether the economic mechanisms that underlie these results conform to what was suggested in Table 6.2 and the rest of the chapter in fact is devoted to a discussion of these issues.

Table 6.4 Fixed-price multipliers giving effects of alternative sugar processing technologies on household income distribution

Households	*Traditional technology*	*Modern technology*
Agricultural employees	0.192	0.110
Farm size 1	0.342	0.205
Farm size 2	0.235	0.148
Farm size 3	0.391	0.239
Rural lower	0.405	0.246
Rural middle	0.065	0.042
Rural higher	0.138	0.094
Urban lower	0.239	0.301
Urban middle	0.038	0.042
Urban higher	0.229	0.244
Total household income	2.275	1.666

Source: Khan and Thorbecke (1988).

Direct and indirect employment effects on household groups

Our earlier analysis of technological systems led us to conclude that direct and indirect employment effects would tend to be greater when traditional rather than modern technologies are used and we further suggested that in the former case the effects would tend to occur in rural rather than urban areas. Table 6.5 throws a good deal of light on these issues because it contains estimates of total employment effects for 16 labour groups, when output of sugar is increased (by 1 million rupiahs) on the basis of both traditional and modern technologies.

Table 6.5 indicates that total employment is indeed greater when the traditional technique is used and it also reveals, as hypothesized, that the employment gains would be greatest in the rural labour categories. Among these, the categories of unpaid rural, paid rural and production unpaid rural feature especially prominently and it can be shown (from an analysis of fixed price multipliers), that much of the income that accrues to these groups is distributed ultimately to the very household groups (such as rural lower) that were seen earlier to gain most from the use of traditional technology.[13] Thus, to at least some extent, the differential impact of alternative technologies on the distribution of income is attributable to the varying amounts of employment with which those technologies are associated.

Table 6.5 Total employment – effects of alternative technologies in the sugar industry

Labour category	*Effects on labour income of increase of 1 million rupiahs in output of* brown sugar (fixed-price multiplier)	white sugar (fixed-price multiplier)	*Average wages and salaries per worker equivalent (thous. rupiahs)*	*Employment (number of worker equivalents)* Brown	White
Ag. paid rural	133	79	92.1	1.44	1.17
Ag. paid urban	5	3	101.0	0.05	0.03
Ag. unpaid rural	218	142	88.1	2.47	1.61
Ag. unpaid urban	9	6	113.5	0.08	0.05
Production paid rural	118	97	82.9	1.42	1.17
Production paid urban	49	69	121.3	0.40	0.57
Production unpaid rural	110	66	76.8	1.43	0.86
Production unpaid urban	17	18	–		
Cler. paid rural	45	35	106.9	0.42	0.33
Cler. paid urban	61	59	191.1	0.32	0.31
Cler. unpaid rural	85	62	–		
Cler. unpaid urban	46	35	–		
Prof. paid rural	45	35	132.1	0.34	0.26
Prof. paid urban	57	51	723.3	0.08	0.07
Prof. unpaid rural	7	6	78.0	0.09	0.08
Prof. unpaid urban	21	18	–		
Total				8.54	6.51

Source: Khan and Thorbecke (1988) and Central Bureau of Statistics, *Social Accounting Matrix Indonesia, 1975*, vol. 1982.

Consumption linkages

Consumption linkage refers to the extent to which income generated in a technological system is spent on goods emanating from within the same system.[14] Would, for example, the rural employment gains from using traditional technology that have just been mentioned, have the further favourable effect for the rural areas of stimulating the demand for products made in those same areas? Basing our argument partly on the Marxist notion of a mode of production, we suggested earlier that this type of linkage would indeed tend to be a potent one in both traditional and modern technological systems in developing countries (that is, that households in both systems would tend to consume a relatively high proportion of products emanating from within their own particular systems).

Consider from this point of view Table 6.6, which contains the marginal expenditure propensities for the products made by traditional and modern technologies in the six dualistic sectors mentioned earlier. Taking rural and urban households as belonging (in an approximate way) to traditional and modern technological systems, respectively, one needs to determine whether the expenditure propensities of the rural group of households are higher for traditional than modern goods and whether the converse applies to the urban households. For some of the product categories in the table, this is clearly the case. Agricultural and rural households, for example, favour traditional over modern cigarettes while urban households tend to exhibit just the opposite pattern of consumption behaviour. Other cases, however, are more ambiguous. The consumption of tea, for example, behaves as expected with respect to rural households, but households in urban areas consume the same amount of both products.

Or again, though all urban groups consume more milled than hand-pounded rice, among rural households there is no single pattern of consumption behaviour. Thus, the data do not provide unqualified support for our view of how income generated within different technological systems tends to be spent.

The economic separation of technological systems

A more general version of the idea underlying consumption linkages is that economic activity *as a whole* tends to occur within rather than between technological systems (though of course there will always be

Table 6.6 Household marginal expenditure propensities in Indonesia, 1975

	Agricultural employees	*Farm size 1*	*Farm size 2*	*Farm size 3*	*Rural lower*	*Rural middle*	*Rural higher*	*Urban lower*	*Urban middle*	*Urban higher*
Traditional sugar	0.006	0.008	0.006	0.004	0.004	0.005	0.004	0.002	0.002	0.001
Modern sugar	0.014	0.013	0.012	0.012	0.012	0.012	0.011	0.009	0.008	0.005
Handpounded rice	0.015	0.064	0.120	0.037	0.020	0.036	0.014	0.004	0.005	0.001
Milled rice	0.175	0.101	0.027	0.033	0.136	0.116	0.72	0.06	0.058	0.012
Farm processed tea	0.003	0.004	0.003	0.002	0.007	0.004	0.002	0.001	0.001	0.001
Processed tea	.001	.001	.001	.001	.001	.001	.001	.001	.001	.001
Dried and salted fish	0.016	0.017	0.016	0.014	0.018	0.020	0.017	0.013	0.012	0.009
Canned fish[1]	0.004	0.003	0.002	0.002	0.003	0.003	0.003	0.002	0.002	0.002
Traditional cigarettes	0.055	0.051	0.035	0.022	0.03	0.009	0.03	0.02	0.013	0.004
White cigarettes	0.00	0.00	0.005	0.009	0.013	0.019	0.008	0.012	0.008	0.010

1. The marginal expenditure propensities for canning and preserving are zero throughout.

Source: Khan and Thorbecke (1988).

Table 6.7 Aggregate effects of alternative sugar technologies on five other dualistic sectors using fixed price multipliers

Technology	*Aggregate effects on traditional dualistic sectors*[a]	*Aggregate effects on modern dualistic sectors*[b]
Traditional	0.197	0.216
Modern	0.129	0.148

Note: a. Includes hand-pounded rice, farm processed tea, dried and salted fish, small-scale canning of fruit and vegetables, clove cigarettes.
b. Includes milled rice, off-farm processed tea, canned fish, medium and large-scale canning of fruit and vegetables, white cigarettes.

Source: Khan and Thorbecke (1988).

some degree of interaction between them). This broader hypothesis can be examined in the context of SAM-TECH by means of the data contained in Table 6.7.

What is shown here is the extent to which a unitary increase in sugar demand (and output) influences economic activity in the traditional as opposed to the modern component of the five other dualistic sectors we have already referred to. Thus, while the traditional sugar processing technology has a much greater impact on the other traditional activities than does the modern technology, the converse does not hold true (for reasons as yet unclear to us).

Structural path analysis of household income generation

Further insights into the mechanisms underlying the results contained in Table 6.3 can be gained by using a technique known as structural path analysis. This technique is particularly well suited to explaining the distributional consequences of using alternative technologies, because it attempts to decompose the fixed price multipliers into a series of both direct and indirect influences.[15] Each average expenditure propensity a_{ji} or marginal expenditure propensity c_{ji} can be interpreted as the magnitude of the influence transmitted from i to j. In our case it is the matrix of marginal expenditure propensities that captures the *direct influence* of the network of elementary paths.

Two complications arise in the transmission of these influences. First, in travelling between successive poles, as the number of poles increase the influence gets attenuated. The attenuation could be calculated simply by multiplying the corresponding MEPs were it not for a further complication. This second complication arises because of the presence of loops and circuits where a set of influences go round and round infinitely many times within a fixed number of poles. It is this second, convergent process which gives rise to the so-called path multipliers. The fixed price multipliers are called global influences. The decomposition of this global influence can now be expressed as follows:

$$\begin{aligned} I^{G}(i \rightarrow j) = m_{ji} &= \sum_{p=1}^{n} I^{T}(i \rightarrow j) \\ &= \sum I^{D}(i \rightarrow j) M_{p} \end{aligned} \tag{8}$$

where $I^{G}(i \rightarrow j)$ = global influence from the ith column in the SAM to the jth row.

$I^{T}(i \rightarrow j)$ = Total influence from i to j.
I^{D} = Direct influence from i to j.
M_{p} = path multiplier along the path p.

The above equation can now be used to analyse the effects of changes in output (in response to increased demand) in either brown sugar or white sugar on the incomes of all the household groups. Since it would be tedious to discuss all 20 influence graphs we have selected three groups for analysis here. (We chose one group to represent farming households, another to represent rural households and a third as representative of households from urban areas.) Figures 6.1, 6.2 and 6.3 capture the structural paths for these groups, respectively. We now turn to a discussion of the influences shown by these particular graphs. (In interpreting them, one should bear in mind that at the origin of each arc the corresponding marginal expenditure propensity is given.)

The first two sets of path diagrams are very similar in at least two major respects. The first is that in both cases the two different technologies exert their influence through the same or very similar sets of paths. In the first figure, that is to say, the paths are identical while in the second figure there is only one extra path (connecting brown sugar to the group 'clerical paid rural') that distinguishes brown from white sugar. The two figures are also very similar with regard to differences in the magnitude of the linkages that apply to the case of brown as opposed to white sugar. In both cases,

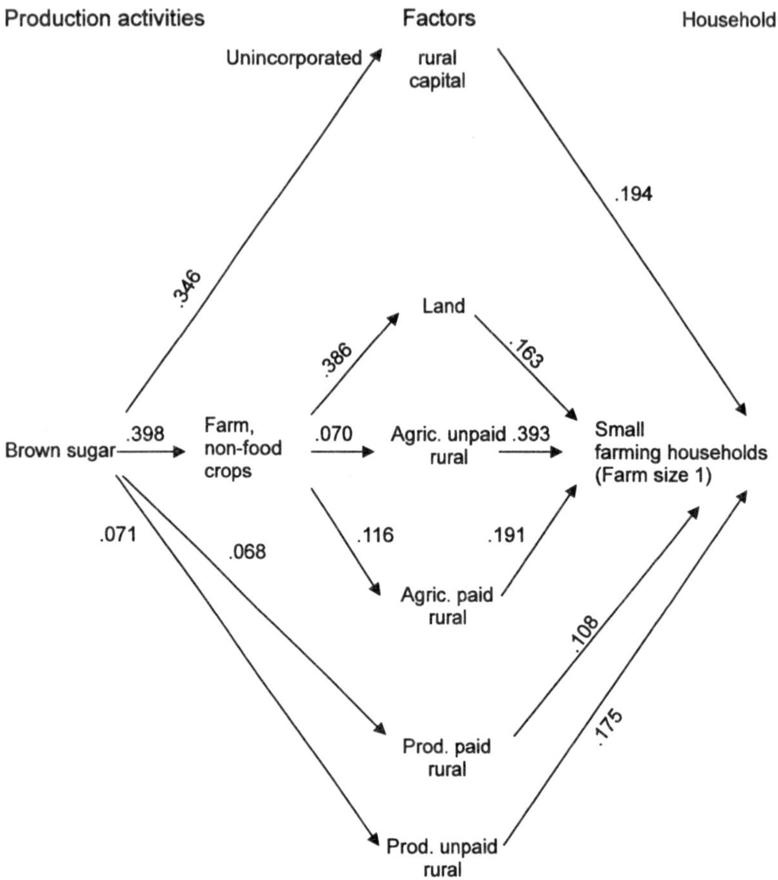

Figure 6.1a Structural path network for the effects of brown sugar on small farming households
Source: Khan and Thorbecke (1988).

that is to say, the main quantitative differences lie in the backward linkage to farm non-food crops (that is, palm sugar and sugar cane for white and brown sugar respectively) and the linkage from sugar (processing) to the group 'unincorporated rural capital'. In each case the magnitude of the linkage is considerably greater for brown than for white sugar. More specifically, whereas the path between brown sugar and non-food crops has a marginal expenditure propensity of 0.398, the corresponding figure for white sugar is 0.186 and whereas the path from brown sugar to unincorporated rural

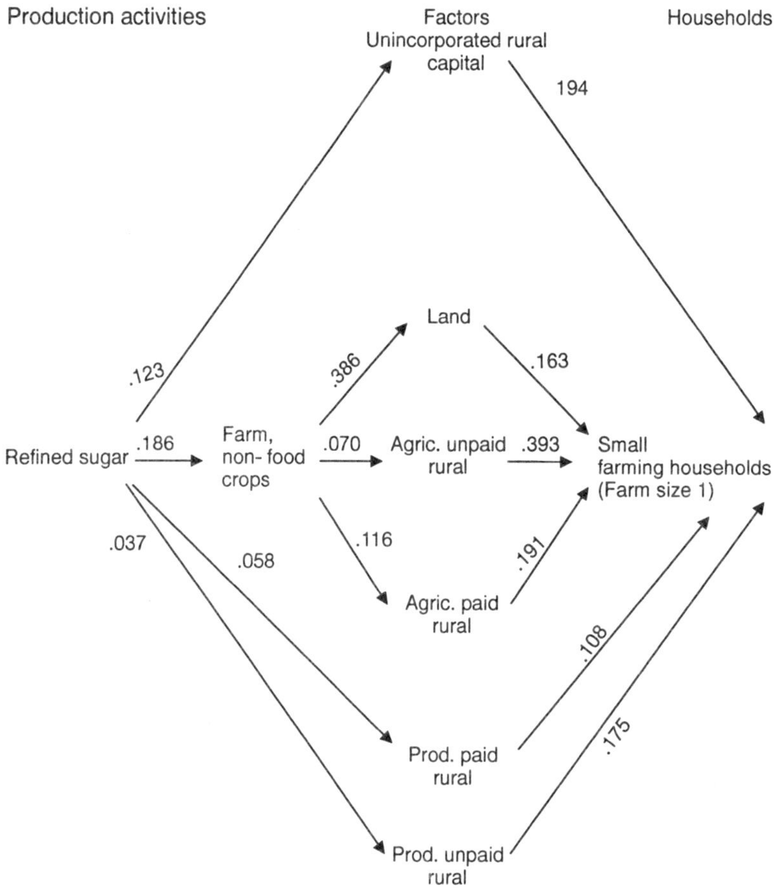

Figure 6.1b Structural path network for the effects of refined sugar on small farming households
Source: Khan and Thorbecke (1988).

capital has a marginal expenditure propensity of 0.346, the same path in the case of white sugar is associated with a marginal expenditure propensity of only 0.123.

Both these differences, we submit, are consistent with what one would expect on the basis of our previous analysis of technological systems. In particular, we suggested there that, for a number of reasons, backward linkage effects would tend to be more pronounced with traditional than with modern techniques. And we also suggested that modes of production in traditional technological

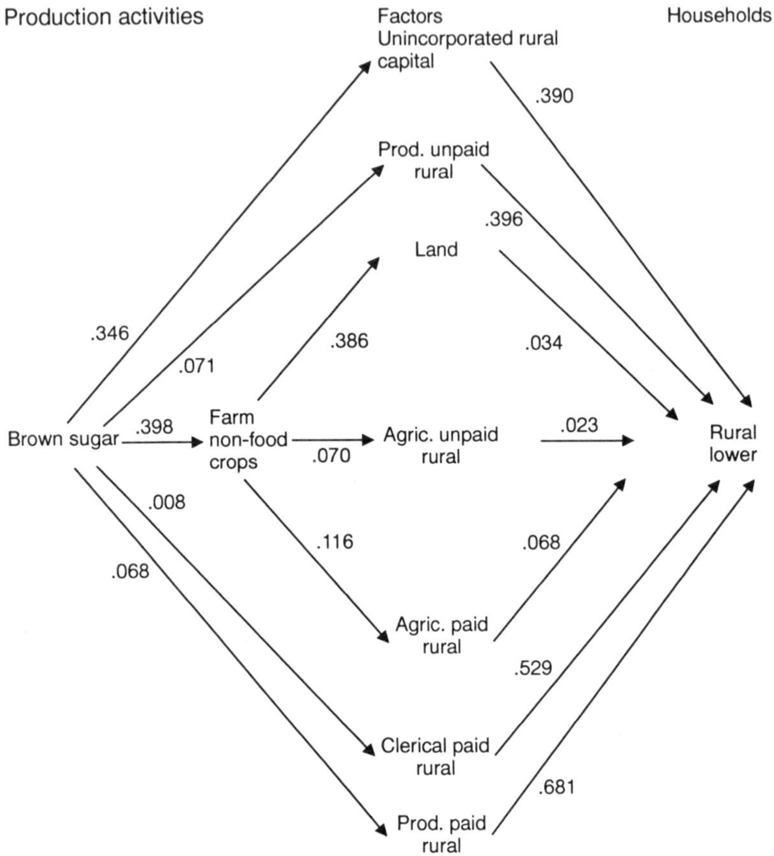

Figure 6.2a Structural path network for the effects of brown sugar on the income of rural lower households
Source: Based on Khan and Thorbecke (1988).

systems are usually such that incomes are spread relatively widely among a large number of small-scale rural producers (who, in terms of SAM-TECH are represented as the owners of 'unincorporated rural capital').

The third path diagram, dealing with 'urban lower households', differs from the two previous paths in that it is the *nature* rather than the *extent* of the linkages that differs between the traditional and modern technologies. Thus, whereas there are three separate paths from refined sugar processing to urban factors of production (and ultimately to 'urban lower households'), no such paths are apparent in the case

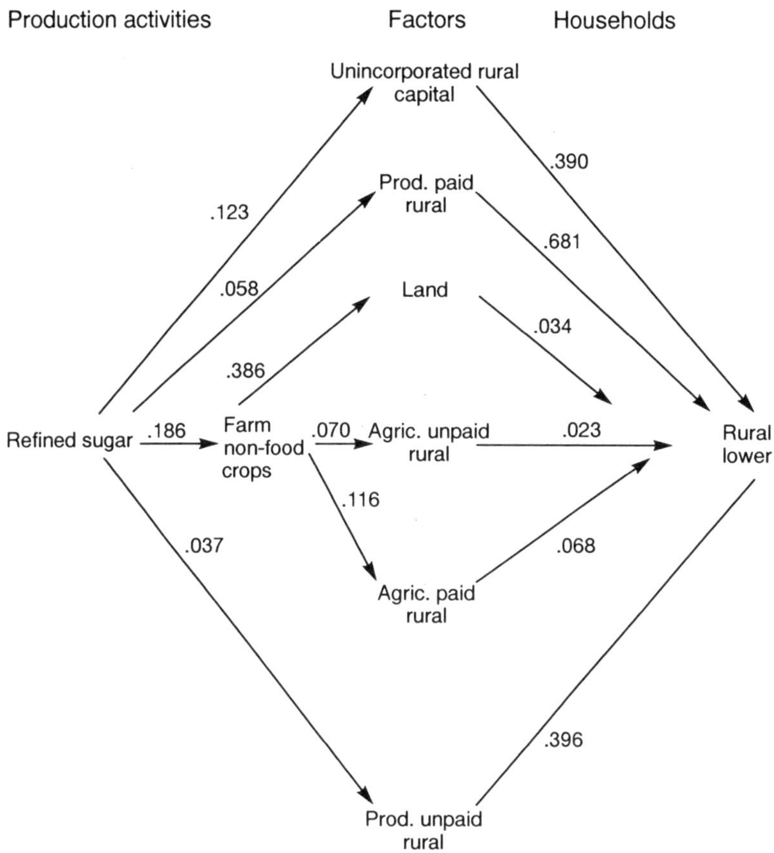

Figure 6.2b Structural path network for the effects of refined sugar on the income of rural lower households
Source: Based on Khan and Thorbecke (1988).

of brown sugar. This rather striking result should be interpreted in conjunction with, and is indeed the mirror image of, the paths linking brown sugar to rural factors of production in Figure 6.1. For, whereas in that case brown sugar exhibited stronger and more numerous linkages to rural factors of production, in the urban context it is (only) refined sugar that exhibits the linkages to factors of production. In short, traditional technologies seem to belong to technological systems that are predominantly rural, whereas modern technologies appear to form part of technological systems that are primarily urban in character.

Figure 6.3a Structural path network for the effects of brown sugar on the income of urban lower households
Source: Based on Khan and Thorbecke (1988).

CONCLUSIONS

Of the many topics that fall under the heading non-agricultural technology and development, the connection between technology choice and income distribution has been treated perhaps the least satisfactorily. The main problem is that this relationship needs to be explored not only, as now, at the micro-economic level, but also at the level of the economy as a whole.

Progression to the latter took the form in our analysis, firstly, of identifying the mechanisms through which technology exerts an influence on the distribution of income at this level. For each of the mechanisms thus identified, we suggested that there are historical reasons why labour-intensive techniques tend to be more egalitarian in their impact than capital-intensive techniques. We were able to test

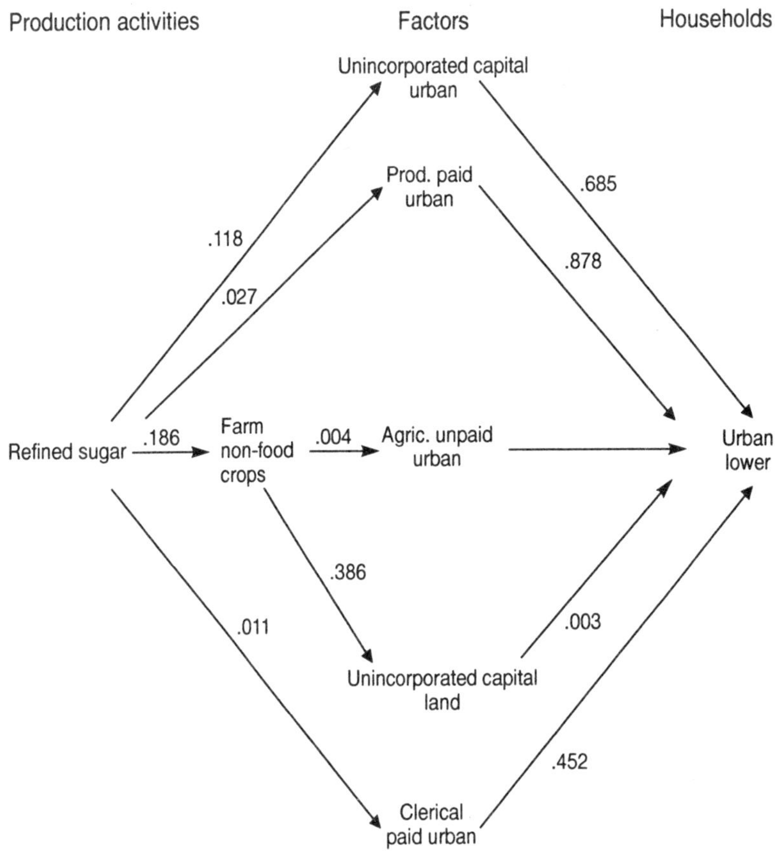

Figure 6.3b Structural path network for the effects of refined sugar on the income of urban lower households
Source: Based on Khan and Thorbecke (1988).

these hypotheses by means of a disaggregated social accounting matrix (SAM) for Indonesia and we found structural path analysis of the SAM to be an especially useful tool for exploring the macro-economic relationships in question. For the most part our empirical results confirm what the theoretical analysis led us to expect.

Notes

1. Mellor (1976).
2. Haggblade and Hazell (1989).
3. Ibid., p. 346.

4. Stewart (1977).
5. Richards (1986).
6. Lange (1963) p. 21.
7. Stewart (1977) p. 7.
8. Ibid., p. 6.
9. Ibid., p. 7.
10. Marx (1946) p. 377.
11. The following discussion is taken from Khan (1989).
12. Khan and Thorbecke (1988, 1989).
13. See Ibid. (1988) for details.
14. For an analysis of consumption linkages see Mellor (1976).
15. See Defourny and Thorbecke (1984); Khan and Thorbecke (1988, 1989) and Khan (1991).

References

J. Defourny and E. Thorbecke (1984), 'Structural path analysis and multiplier decomposition within a social accounting matrix framework', *Economic Journal*, vol. 94, March.

S. Haggblade and P. Hazell (1989), 'Agricultural Technology and Farm – Non-farm Growth Linkages', *Agricultural Economics*, vol. 3.

J. James and H. Khan (1993), 'The employment effects of an income redistribution in developing countries', *World Development*, vol. 21, no.5.

H. Khan (1989), *The Political Economy of Sanctions Against Apartheid*, Boulder: Lynne Rienner.

H. Khan (1991), 'Multiplier and structural path analysis of trade sanctions', in G. Shepherd Jr (ed.) *Effective Sanctions on South Africa*, New York: Greenwood.

H. Khan and E. Thorbecke (1989), 'Macroeconomic effects of technology choice: multiplier and structural path analysis within a SAM framework', *Journal of Policy Modelling*, vol. 11, no.1.

H. Khan and E. Thorbecke (1988), *Macroeconomic Effects and Diffusion of Alternative Technologies within a Social Accounting Matrix Framework*, Aldershot: Gower.

O. Lange (1963), *Political Economy*, Oxford: Pergamon.

K. Marx (1946), *Capital*, London: Allen & Unwin.

J. Mellor (1976), *The New Economics of Growth*, Ithaca: Cornell University Press.

A. Richards (1986), *Development and Modes of Production in Marxian Economics: A Critical Evaluation*, London: Harwood.

F. Stewart (1977), *Technology and Underdevelopment*, London: Macmillan.

APPENDIX TO CHAPTER 6
TECHNOLOGY CHOICE AND INCOME DISTRIBUTION IN OTHER DUALISTIC SECTORS

As already noted, this chapter has dealt with only one out of the six dualistic sectors studied by Khan and Thorbecke. What we have not yet indicated, however, is that sugar is by no means an exception to the overall pattern of influence of alternative technologies on the distribution of income. On the contrary, as we shall now see, Khan and Thorbecke found not only a similar pattern of distributional outcomes across the six dualistic sectors, but also considerable similarity in the factors underlying those outcomes.

Consider first the general conclusion that 'for a developing country such as Indonesia, the traditional technique often has more linkages with the rest of the economy, creates greater employment and benefits especially the rural and low-skilled workers more than the modern alternative.... Therefore, unless the traditional alternative is clearly anachronistic and inefficient, a strong case exists for its use along with the modern technique, at least as far as employment creation, improvement of income distribution and poverty alleviation continue to be major policy objectives.' [1] Table 6A. 1, for example, shows that in the five sectors other than sugar, traditional technology benefits low-skilled, rural labour factors more than the modern alternative.

In particular, with the exception of fruit and vegetable canning, agricultural paid and unpaid rural labourers exhibit higher fixed-price multipliers under traditional than modern techniques, thus replicating the sugar example described above in relation to Table 6.5. And again as in the case of sugar, traditional technologies in the other sectors tend to generate more overall employment than their modern counterparts. This can clearly be seen for example in rice production for the 16 labour factors shown in Table 6A.2. For whereas the production of one million rupiahs of handpounded rice generates 15.8 worker equivalents, the corresponding figure for milled rice is only 12.5.

Generally speaking, moreover, these similarities in sectoral outcomes can be explained in terms of factors that are common to all six dualistic sectors. In particular, 'The favourable impact of traditional as compared to corresponding modern techniques on employment, poverty alleviation...in the *short run* hinges on the greater backward linkages and smaller leakages of these techniques.'[2] With regard to the first of these two explanatory variables, we have already seen that traditional sugar technology exhibits greater backward linkages than the production of sugar by modern methods and that this differential effect, in turn, primarily benefits low-skilled, rural labour. A similar description applies to rice production, inasmuch as 'the traditional technology requires relatively more paddy inputs per unit of output than the corresponding modern technology. In other

Table 6A.1 Rural low-income generation and alternative technologies*

	Rice		*Cigarettes*		*Fish*		*Tea*		*Canning*	
	Hand-pounded	*Milled*	*Clove*	*White*	*Dried and salted*	*Canned*	*Farm Processed*	*Processed*	*Traditional small-scale*	*Modern large-scale*
Agric. paid rural	0.252	0.169	0.093	0.045	0.109	0.099	0.254	0.120	0.099	0.133
Agric unpaid rural	0.624	0.437	0.142	0.080	0.194	0.179	0.401	0.168	0.244	0.281

**Note:* All figures represent fixed price mulitipliers.

Source: Khan and Thorbecke (1988).

Table 6A.2 Total employment effects of alternative technologies in the rice industry

	Effects on labour income of increase of 1 million rupiahs in output of		*Average wages and salaries per worker equivalent rupiahs 000*	*Employment (Number of worker equivalents)*	
Labour category	*Hand-pounded rice (rupiahs 000)*	*Milled rice (rupiahs 000)*		*Hand-pounded rice*	*Milled rice*
Agric. paid rural	252	169	75.9	3.32	2.23
Agric. paid urban	7	6	98.5	0.07	0.06
Agric. unpaid rural	624	437	71.2	8.76	6.14
Agric. unpaid urban	18	17	89.4	0.20	0.19
Prod. paid rural	48	70	126.3	0.38	0.55
Prod. paid urban	47	55	189.7	0.25	0.29
Prod. unpaid rural	38	62	102.1	0.37	0.61
Prod. unpaid urban	16	16	160.0	0.10	0.10
Clerical paid rural	34	35	111.6	0.30	0.31
Clerical paid urban	59	60	198.9	0.30	0.30
Clerical unpaid rural	81	78	80.3	1.01	0.97
Clerical unpaid urban	45	43	117.7	0.38	0.37
Prof. paid rural	41	39	349.2	0.12	0.11
Prof. paid urban	57	56	556.3	0.10	0.10
Prof. unpaid rural	8	7	101.0	0.08	0.07
Prof. unpaid urban	21	20	419.1	0.05	0.05
			Total	15.79	12.45

Source: Khan and Thorbecke (1988).

words, the linkage of the former product-cum-activity to paddy production is greater than the latter which, in contrast, requires more *nonagricultural* intermediate inputs per unit of output.'[3]

Systematic differences in leakages between techniques, the second of the two explanatory variables, helps to explain the similar distributional outcomes across sectors in precisely the manner that one would expect on the basis of differences between traditional and modern technological systems. In particular, as predicted in this chapter, 'Traditional production methods are generally less import-intensive both directly (in terms of intermediate import requirements) and indirectly . . . the resulting income distribution among socioeconomic groups is tilted more towards those groups which spend less on finished imports and import-intensive domestic goods.'[4]

Notes

1. Khan and Thorbecke (1988) p. 170.
2. Ibid., p. 169.
3. Ibid., p. 85, emphasis added.
4. Ibid., p. 169.

References

H. Khan and E. Thorbecke (1988). *Macroeconomic Effects and Diffusion of Alternative Technologies within a Social Accounting Matrix Framework*, Aldershot: Gower.

Conclusions

This volume has been organized around the notion of a technological system, and more specifically the distinction between modern and traditional technological systems. In our treatment of the subject we have been able to find a way of identifying distinct systems of technology in a multisectoral model of the economy. This approach has also allowed us to quantify certain aspects of different technological systems within a rigorous conceptual framework. As already noted in the introduction, the concept we have used has historical connections with the idea of modes of production – both pre-capitalist and capitalist. Our approach has been to identify different technology systems, study their characteristics and understand their sectoral and economy-wide impacts. We have argued that although technological systems are rooted in these different historical periods they can nonetheless still be identified as distinctive entities in the context of contemporary developing countries.

The process we have in mind can be illustrated particularly clearly with reference to the historical evolution of the manufacturing sector in Nigeria, where, before the turn of the last century, extensive markets for locally produced artisan goods are known to have existed. For example,

> In tobacco production, markets for locally produced tobacco products such as .... 'bookies' date at least as early as 1900..... Bookies are homemade cigarettes produced from local tobacco and derive their name from the 'holy Book' whose pages originally provided the cigarette paper. As early as 1860 local brewing industries produced beer from local corn and millet and strong wine from kola nuts in many parts of the country. Cotton had been grown and manufactured in Nigeria for many centuries: 'Spun, hand-woven in simple cloth, and dyed with colours obtained from native plants, it provided most of the clothing of the people. Nearly all the ginning, spinning and weaving equipment was made of wood.' Building materials, particularly laterite mud compounds, were used long before cement products were imported. And the capabilities for processing timber products and manufacturing unrefined sugar products from cane existed long before the arrival of their modern competitors.[1]

These examples well illustrate the notion of a traditional technological system in which small-scale, artisanal producers were using local inputs to supply relatively simple domestic consumption needs. At a much later stage this system was threatened by the entry into Nigeria of multinationals producing similar, but more sophisticated, products on the basis, mainly, of imported rather than local technologies and factors (it was threatened, that is, by the emergence of a modern technological system). The traditional system, however, was only partly displaced by its modern counterpart, to a great extent, it seems, because large numbers of Nigerians continued to live at subsistence levels of income and to demand the types of product that were formerly associated with those incomes. 'This phenomenon of market segmentation', writes Peter Kilby,

> flows from the low level of per capita income. With severely limited purchasing power, the community's poorest members will often choose inexpensive low-quality products such as tyre-sandals or household utensils beaten out of scrap by the tinker in preference to the higher quality, more durable, dearer alternative.[2]

In large part, therefore, the coexistence of divergent technological systems in contemporary developing countries can be ascribed to the close interactions between different groups of users and producers of technologies, final products and inputs. It is worth noting that a similar argument also underlies the more aggregate notion of a *national* technological system that has been advanced in the very different context of comparisons between developed countries. Noting that technologies often seem to have a distinct national 'flavour', for example, some authors view national systems as products of a historical process that cannot easily be transferred between countries and they also emphasize the importance for technology of intense user–producer interactions within national frontiers. Thus: 'Interaction between users and producers belonging to the same national system may work efficiently for several reasons. Short geographical distance is part of the explanation; more important may be a common language and the cultural proximity.'[3] Conversely: 'According to the user producer approach, geographical and cultural distance is a factor which may impede the interaction between user and producers. This might contribute to an explanation of why different national systems display different patterns of development.'[4]

As with our own analysis of dualistic technology systems within developing countries, the notion of a national technological system entails a rather marked departure from the neoclassical tradition in economics. For whereas both concepts of a technological system rely on history, culture and geography, neoclassical analysis tends on the whole to regard such factors as lying outside its purview.

TECHNOLOGY, EMPLOYMENT AND INCOME DISTRIBUTION

The previous chapters have mainly been concerned to use the concept of a technological system for the analysis of two distinct but related issues. The first is the manner in which the distribution of income, through its impact on patterns of demand, influences technology and employment, and the second is how the distribution of income is itself influenced by the choice of technology.[5] These questions have rarely been posed in the choice of technology literature, mainly because that literature, based as it usually is on neoclassical micro-economics, tends to take variables other than factor prices as given. As can readily be seen in most textbooks, therefore, the only relevant policy intervention is one of influencing the relative prices of capital and labour (that is, of influencing the shape of the iso-cost line and hence the point of tangency of that line and an isoquant). Yet, as Stewart rightly points out, 'income distribution is both part cause and product of technical choice'[6] and hence the latter may be influenced by policy towards the former as well as vice versa.

As noted in Chapter 1, we used both micro and macro methods to analyse these issues on the basis of far more disaggregated data about products, techniques and households than is usually available. In what follows we summarize the results of our enquiries and show how they can be interpreted from the standpoint of technological systems. Where relevant, moreover, we describe the significance of the overall findings for major development policy issues.

Income redistribution, technology and employment

It has been argued by some authors that there is no reason to expect the systematic association between products, processes and incomes

Table 7.1 Case studies of products, processes and incomes

Sector	*Country*	*Chapter*	*Products compared*
Textiles	India	2	Hand-loom and mill-made clothing of different counts
Sweeteners	India	3	Refined and unrefined products
Soap	India	2	Laundry soaps made by firms of different sizes
Furniture	Kenya	2	Household furniture of varying quality
Building materials	Kenya	1	Mud, wattle, timber, cement blocks, etc.
Roof-tiles	Indonesia	2	Open fire, hand-pressed and power-pressed tiles
Soap	Bangladesh	2	Laundry and toilet soaps made by different methods
Soap	Ghana	2	Laundry and toilet soaps made by different methods
Manufactures	Nigeria	2	Varieties of furniture, shoes and soap

that is associated with the notion of coexisting technological systems in developing countries, namely, that relatively poor (rural) households consume more labour-intensive variants of given manufactured products than their high-income (urban-based) counterparts.[7] Yet, several of the previous chapters indicate that over a range of different products and countries the postulated relationship does indeed appear to hold (see the summary contained in Table 7.1).

In each of the cases reported in Table 7.1, a redistribution of income in favour of the poor would increase employment in a partial equilibrium sense, because the composition of sectoral output is tilted towards the relatively labour-intensive product. Consider, for example, furniture-making in Kenya, which is one of the cases shown in Table 7.1. A series of simulations conducted for this industry by William House reveal that such an income redistribution would cause 'a major increase in employment in the furniture industry',[8] that 'most of these additional jobs would be created in the rural and urban informal sectors, where the average earnings are lowest'[9] and that 'The more severe the income redistribution taking place the greater would be the growth of lower-income employment.'[10]

In most studies of the employment effects of an income redistribution, however, this particular mechanism would not be captured because products tend to be rather broadly defined (in categories such as furniture, soap and textiles) and hence make no distinctions of the kind found in the last column of Table 7.1.[11] This in turn means that such studies tend to underestimate the actual employment effect of an income redistribution in favour of the poor, because they fail to consider that *within* the product categories thus defined, the low-income groups tend to consume the relatively labour-intensive alternatives. The extent of this aggregation bias is examined in Chapter 3 on the basis of a case study of the Indian sugar processing industry. Aggregation bias is defined as the difference between aggregated and disaggregated estimates of the employment effect of an income redistribution and in the case in question it is shown to be of the order of 50 per cent. Chapter 5 is also concerned with measuring the extent of aggregation bias but in a more complete estimation framework which takes into account indirect as well as direct employment effects. Chapter 5, that is to say, examines the implications of disaggregation not just for the direct employment effects of an income redistribution in favour of the poor but also the indirect or linkage effects. Using a disaggregated social accounting matrix (SAM) for this purpose, we again compared the results of an income redistribution based on aggregated as opposed to disaggregated data and we found that the former underestimates the actual employment effect by a factor of around three. Our analysis of the sources of this discrepancy indicated that they were the result in part of a tendency for aggregation to understate linkage effects most markedly among relatively unskilled rural labourers. We further suggested that this tendency is consistent with the implications of the distinction between traditional and modern technological systems in developing countries.

In thus quantifying the implications for technology and employment of alternative distributions of income, we are in effect reviving the importance of demand in the choice of development strategies. More specifically, for example, our results bear directly on issues that were raised more than twenty years ago by Johnston and Kilby, who then argued in favour of a 'unimodal' development strategy which raises the productivity of a large fraction of a country's farming units, as opposed to a 'bimodal' strategy which concentrates output among a small subsector of farming units.[12] Their preference for the former was based partly on the differential demand-side effects that were thought to be associated with the two strategies. On the one hand, a relatively egalitarian distribution of income would lead, as they saw it, to a demand for comparatively simple goods that could be manufactured with rudimentary technological capabilities largely on the basis of local intermediate inputs. An unequal distribution of income, by contrast, would lead to a limited market for sophisticated products and large-scale inappropriate technologies. At the time, Johnston and Kilby were able to find only one data-set (for India) with the required degree of disaggregation to test these propositions. The body of evidence that we have compiled in this book, however, considerably strengthens their original case for a 'unimodal' rather than a 'bimodal' strategy of development.

Technology choice, employment and income distribution

Whereas the previous results refer to the manner in which the distribution of income, through its impact on patterns of demand, influences technology and employment, Chapter 6 tackles the causality that runs between these variables from the opposite direction. That is, it is concerned with the question of how the distribution of income is itself influenced by the choice of technology. Chapter 6 nevertheless bears a close affinity to the previous chapter in that it is also based conceptually on the notion of a technological system and because, as in that chapter, the analysis is conducted at the macro-economic level using the same disaggregated SAM as before.

The notion of a technological system – and more specifically the notion of coexisting but divergent technological systems – is used to derive an initial set of hypotheses concerning the differential employment and income distribution effects of using traditional as opposed to modern methods of production. Among the most important of the hypotheses that derive from this analytical framework is that, on account of their more intensive backward linkage effects and lesser propensity to generate leakages (via for example imported inputs), the former technologies tend to generate more aggregate employment than the latter. Such additional employment, we further suggested, tends to occur mainly among relatively unskilled and low-income workers, thus exerting an egalitarian influence on the household distribution of income. These hypotheses were examined not only in relation to a disaggregated SAM, as already noted, but also with the aid of structural path analysis, which is an especially useful tool for decomposing the ultimate distributional effects of alternative technologies into their constituent elements.

Although the results based on these methods do not lend unqualified support to each and every one of the hypotheses that we advanced at the outset of Chapter 6, they do confirm our main expectations, firstly, because 'the traditional technique often has more linkages with the rest of the economy, creates greater employment and benefits especially the rural and low-skilled workers more than the modern alternative'[13] and secondly, 'Traditional production methods are generally less import-intensive both directly (in terms of intermediate import requirements) and indirectly ... the resulting income distribution among socioeconomic groups is tilted more towards those groups which spend less on finished imports and import-intensive domestic goods'.[14]

These findings, we believe, make a methodological contribution not only to the debate on appropriate technology, which hitherto has been conducted almost entirely at the micro level, but also to the case for appropriate technology itself. For what is important to policy-makers is not simply the direct economic effects of alternative technological choices but also – and indeed often more importantly – the indirect or feedback effects, which in our model distinctly favour the traditional over the modern alternative, at least for an economy with the need to generate employment and income for the poor.

OVERALL CONCLUSIONS

The findings presented in this volume concerning the relationships between technology, employment and income distribution have a distinctly symmetrical character. On the one hand, that is, a relatively equal initial distribution of income generates a pattern of demand that favours labour-intensive technologies and hence employment, while those technologies themselves contribute towards equality in the size distribution of income. Conversely, a highly inegalitarian pattern of income distribution tends to be associated with imported, capital-intensive technologies, which in turn reinforce or accentuate the initial inequality. These symmetrical patterns, we have argued, are not simply a coincidence but are rather a result of the fact that in most developing countries one can identify different technological systems, each with its own distinguishing characteristics.

In actual country practice, of course, depending, among other factors, on government policy, either of these two mutually reinforcing sets of relationships between technology, employment and income distribution may occur. The first possibility is perhaps best illustrated by the case of Taiwan, where, on the one hand, appropriate technology was extensively used in industry[15] and where, on the other hand, the distribution of income in the early stages of growth was relatively equal, thus providing a local market for the products typically associated with appropriate industrial technologies.[16] In most other contemporary developing countries, by contrast, the outcome is just the opposite in that an unequal distribution of income conduces to and reinforces the preponderance of inappropriate technology in the industrial sector.[17]

Notes

1. Biersteker (1978) p. 105.
2. Kilby (1969) p. 33.
3. Lundvall (1988) p. 360.
4. Ibid.
5. Although Chapter 4 does not concern itself with either of these two major issues, it does nevertheless draw heavily on the notion of coexisting technological systems. For what the chapter shows is that products are designed to form part of particular technological systems and the welfare effects of consuming those same products in a different technological system may vary considerably. More specifically, we show that new medicinal drugs are designed to form part of modern (developed country) technological systems, and where elements of those systems are missing the likely outcome is product misuse and a lesser contribution to individual health functionings.
6. Stewart (1975) p. 13.
7. See for example Cooper (1990).
8. House (1984) p. 164.
9. Ibid., p. 162.
10. Ibid.
11. Morawetz (1974).
12. Johnston and Kilby (1975).
13. Khan and Thorbecke (1988) p. 170.
14. Ibid., p. 169.
15. Ranis (1973).
16. Johnston and Kilby (1975).
17. See Stewart (1977) for a description of these self-reinforcing cycles.

References

T. Biersteker (1978). *Distortion or Development? Contending Perspectives on the Multinational Corporation*, Cambridge, Mass.: MIT Press.

C. Cooper (1990). 'Notes on Income Distribution, Growth and Demand', in F. Stewart, H. Thomas and T. de Wilde (eds), *The Other Policy*, London: IT Publications.

W. House (1984). 'Product Choice and Employment in Furniture-Making in Kenya', in W. van Ginneken and C. Baron (eds) *Appropriate Products, Employment and Technology*, London: Macmillan.

B. Johnston and P. Kilby (1975). *Agriculture and Structural Transformation*, Ithaca: Cornell University Press.

H. Khan and E. Thorbecke (1988). *Macroeconomic Effects and Diffusion of Alternative Technologies within a Social Accounting Matrix Framework*, Aldershot: Gower.

P. Kilby (1969). *Industrialization in an Open Economy: Nigeria 1945–1966*, Cambridge University Press.

B. Lundvall (1988). 'Innovation as an Interactive Process: from User-Producer Interaction to the National System of Innovation', in G. Dosi et al. (eds). *Technical Change and Economic Theory*, London: Frances Pinter.

D. Morawetz (1974). 'Employment Implications of Industrialisation in Developing Countries: A Survey', *Economic Journal*, vol. 84, September.

G. Ranis (1973). 'Industrial Sector Labor Absorption', *Economic Development and Cultural Change*, vol. 21, No.3.

F. Stewart (1977). *Technology and Underdevelopment*, London: Macmillan.

Index